Thermoelectric Materials and Devices for Clean Energy Harvesting

This book is designed to demonstrate the need for thermoelectric energy materials, including non-conventional materials for heating and cooling applications in thermoelectric energy-efficient devices. In addition to the fundamentals, functionalities, and classifications associated with thermoelectric energy materials that can be used in industries, this book explores how these materials can contribute to high energy efficiency and sustainability, even though they are used in waste heat recovery systems. Climate change and global warming are also taken into consideration.

Features:

- Focuses exclusively on non-conventional thermoelectric materials.
- Deals with key areas for new generation thermoelectric materials, such as nanostructured materials and non-traditional materials.
- Highlights the right thermoelectric material selection, their design, characterization, and testing.
- Covers additive manufacturing/3D printing for highly customized and efficient thermoelectric devices.
- Includes case studies based on real-world applications.

This book is aimed at researchers and graduate students in materials science and clean technologies.

Advanced Materials Processing and Manufacturing

Series Editor: Kapil Gupta

The CRC Press Series in *Advanced Materials Processing and Manufacturing* covers the complete spectrum of materials and manufacturing technology, including fundamental principles, theoretical background, and advancements. Considering the accelerated importance of advances for producing quality products for a wide range of applications, the titles in this series reflect the state-of-the-art in understanding and engineering the materials processing and manufacturing operations. Technological advancements for enhancement of product quality, process productivity, and sustainability are on special focus including processing for all materials and novel processes. This series aims to foster knowledge enrichment on conventional and modern machining processes. Micro-manufacturing technologies such as micro-machining, micro-forming, and micro-joining, and hybrid manufacturing, additive manufacturing, near net shape manufacturing, and ultra-precision finishing techniques are also covered.

Nanofinishing of Materials for Advanced Industrial Applications
Edited by Faiz Iqbal, Dilshad Ahmad Khan, and Zafar Alam

Environmentally Benign Machining
Edited by Şenol Bayraktar and Sunil Pathak

Post-Processing of Parts and Components Fabricated by Fused Deposition Modeling Techniques and Advancements
Edited by Vinayak R. Malik, Vivek Tiwary, and Arunkumar P.

Forming and Machining of Polymers, Ceramics, and Composites
Edited by Matruprasad Rout and Kishore Debnath

Fundamentals and Advances in Metal Matrix Composites
Edited by Tharmaraj Ramakrishnan and PM Gopal

Thermoelectric Materials and Devices for Clean Energy Harvesting Properties, Fabrication, and Opportunities
Ram Krishna and A. D. Dhass

For more information about this series, please visit: www.routledge.com/Advanced-Materials-Processing-and-Manufacturing/book-series/CRCAMPM

Thermoelectric Materials and Devices for Clean Energy Harvesting

Properties, Fabrication, and Opportunities

Ram Krishna and A. D. Dhass

CRC Press
Taylor & Francis Group
Boca Raton London New York

CRC Press is an imprint of the
Taylor & Francis Group, an **informa** business

Designed cover image: Shutterstock

First edition published 2025
by CRC Press
2385 NW Executive Center Drive, Suite 320, Boca Raton FL 33431

and by CRC Press
4 Park Square, Milton Park, Abingdon, Oxon, OX14 4RN

CRC Press is an imprint of Taylor & Francis Group, LLC

ISBN: 9781032619323 (hbk)
ISBN: 9781041072096 (pbk)
ISBN: 9781003639343 (ebk)

DOI: 10.1201/9781003639343

Typeset in Times
by codeMantra

Ram Krishna: *This book is dedicated to all the pioneers and visionaries in the field of thermoelectric materials and devices, whose relentless pursuit of knowledge and innovation has paved the way for cleaner and more sustainable energy solutions.*

To the researchers, scientists, and engineers whose dedication and hard work have contributed to the advancement of thermoelectric technology, enabling the harnessing of waste heat and converting it into usable energy, thereby addressing some of the most pressing environmental challenges of our time.

To the educators and mentors who have inspired and guided new generations of scientists and engineers, fostering a spirit of curiosity, resilience, and excellence in pursuing knowledge.

And finally, to the future generations of innovators who will continue to explore, discover, and develop new ways to harness clean energy, ensuring a sustainable and prosperous future for all.

May this work serve as a source of inspiration and a foundation for continued progress in the quest for clean energy harvesting through thermoelectric materials and devices.

A.D. Dhass: *I would like to thank my professors and supervisors; Dr. P Lakshmi, Head of the Department, Department of Electrical and Electronics Engineering; Former Dean Dr. E Natarajan (Late), College of Engineering, Guindy, Anna University, Chennai, India; and Mr M Jagannathan, Cheyyar, India, for their never-ending love and support that allowed me to complete this book successfully.*

Contents

Preface

This book, *Thermoelectric Materials and Devices for Clean Energy Harvesting: Properties, Fabrication, and Opportunities,* aims to demonstrate the need for thermoelectric energy materials other than conventional materials for heating and cooling applications in thermoelectric devices. Importantly, non-conventional thermoelectric materials are still in the research and development stage, and their high cost and low scalability limit their commercialization. Despite this, their high efficiency and sustainability make them promising candidates for future thermoelectric applications.

This book consists of sections such as an overview of conventional and non-conventional thermoelectric energy materials and their importance to waste heat recovery systems as a potential source of high energy efficiency and sustainability. It encompasses a discussion on the fundamentals, functionalities, and classifications of thermoelectric energy materials that can be used in industry.

Further, several methods are discussed to improve the efficiency and effectiveness of thermoelectric devices by optimizing the process. In addition, a thorough discussion is provided on designing and fabricating thermoelectric materials and devices, as well as heating, cooling, and performing analyses of thermoelectric generators and materials.

Various thermoelectric materials and devices that can be printed using additive manufacturing techniques are also investigated. Moreover, more innovative and practical applications of thermoelectric devices in different fields, such as environmental monitoring, industrial waste heat recovery, medical implants, wearable technology, automotive waste heat recovery, and spacecraft power generation, are covered.

As a critical component in thermoelectric devices, thermoelectric materials' key properties and structures that affect thermoelectric performance are investigated. Challenges and opportunities in the industrial application of thermoelectric devices include finding materials that simultaneously exhibit high electrical conductivity, low thermal conductivity, and high Seebeck coefficient, which are also highlighted.

Additionally, the cost of materials and fabrication processes for thermoelectric devices remains high, limiting their widespread commercialization. In the end, a discussion is provided on ongoing research and development in thermoelectric materials and device design, which are expected to overcome these challenges and pave the way for future research and innovations by scholars, engineers, and researchers.

Acknowledgments

Ram Krishna: This book, *Thermoelectric Materials and Devices for Clean Energy Harvesting: Properties, Fabrication, and Opportunities*, results from many individuals and institutions' collaborative efforts and support. First, I sincerely thank the National Institute of Technology (NIT) Jamshedpur for providing an excellent environment and resources. Special thanks go to my colleagues in the Department of Metallurgical and Materials Engineering for their continuous support and encouragement. I am deeply grateful to my mentors and advisors, whose guidance has been instrumental in shaping my research journey. Their invaluable insights and constructive feedback have significantly enriched this work. I also sincerely thank the University of Leicester, United Kingdom, where I completed my PhD. The knowledge and experience gained there have been foundational to my research in advanced materials. I would like to acknowledge the contributions of my research collaborators and co-authors. Their dedication, expertise, and collaborative spirit have been vital in the successful completion of this book.

A special thank you to my students and research assistants, whose hard work and enthusiasm have inspired me. Finally, I am profoundly grateful to my family and friends for their unwavering support and understanding throughout this journey. Their love and encouragement have been my greatest strength. Thank you all for your invaluable contributions and support. This book is a testament to our collective efforts and shared vision for a sustainable future through clean energy technologies.

A. D. Dhass thankfully acknowledges the support provided by the Indus University, Rancharda, Ahmedabad, Gujarat, India.

Author Biographies

Ram Krishna is an assistant professor in the Department of Metallurgical and Materials Engineering at the National Institute of Technology (NIT) Jamshedpur. He has a robust educational background specializing in metallurgical and materials engineering and holds a PhD from the University of Leicester, United Kingdom. His research interests include advanced materials processing, multiscale characterization of materials, and the development and characterization of new materials for energy technology applications. His extensive research work and publications have significantly contributed to the field of materials engineering.

A.D. Dhass received his BE, ME, and PhD degrees in electrical and electronics engineering, energy engineering, and solar energy in 2005, 2009, and 2018, respectively, from Anna University, Chennai, India. Since 2020, he has been working as an assistant professor at the Indus Institute of Technology and Engineering (IITE), Indus University, Ahmedabad, Gujarat, India. His areas of research include solar energy, thermoelectric generator materials, heat transfer, and phase change materials.

1 Fundamentals of Thermoelectric Generators

Principles, Materials, and Sustainability

1.1 INTRODUCTION

Sustainable energy and innovative renewable energy are global priorities. From nanoWatts (nW) to hundreds of milliWatts (mW), harvested energy operates small- and medium-sized computers and electrical devices without the need for primary energy source. Energy harvesting captures lost energy, stores it, and uses it in micro- and nanoelectronic systems. Photoelectric, electromagnetic, piezoelectric, and radio-frequency are some of the energy harvesting sources. The ability of thermoelectric generators (TEGs) to convert heat and electricity, their huge energy potential, and their diverse applications have made them effective energy harvesters. Reusing environmental heat makes TEGs a leading renewable energy source [1].

Many manufacturing processes recover waste heat. In 2016, 72% of global electrical energy generation came from industrial and chemical waste heat, 63% of which was low-grade. Thermoelectric generator (TEG) thermal energy harvesting has been intensively explored due to its simplified structure, lack of moving parts or vibration, continuous solid-state quiet operation for many years, good scalability, no chemical reaction, and no harmful residuals. Thermal energy harvesting generates electricity using thermoelectric or pyroelectric effects.

The cooling towers lose two-thirds of the power station's electricity as waste heat because gas or steam-powered turbines provide most electrical power by burning fuel to generate heat. The turbine then converts heat into mechanical energy and the generator converts it into electrical energy. Thus, power plant gearbox cables receive only one-third of the fuel energy. Using waste heat from these operations to generate electricity would greatly improve power generation efficiency. The environment benefits from less wastage because less fuel is burned to produce the same amount of power. Thus, TE-based energy conversion systems can tackle the global energy crisis and boost sustainability. TE materials lower greenhouse gas emissions and boost renewable energy by efficiently converting waste heat into power [2].

DOI: 10.1201/9781003639343-1

Thermoelectric devices are characterized by their small size, lack of noise, durability, and user-friendly nature, all of which can be attributed to the absence of any mechanical components. These dependable devices have been utilized in sensors for medical applications and as glazing for energy-efficient structures.

Societies need power, and waste heat matters too; this can be possible through the use of TEGs. By coupling more pairs of TE couples electrically in series and thermally in parallel, each TE module converts thermal energy into electricity [3].

1.2 THERMOELECTRIC GENERATORS

The temperature differential between two materials directly generates energy in thermoelectric (TE) phenomenon. In 1821, Thomas Seebeck identified the effect when he heated a wire connection to demonstrate electromotive force. Figure 1.1 illustrates the Seebeck effect in a thermoelectric unicouple, where a voltage difference occurs at the open ends due to fluctuations in the temperature gradient and the pathway for free charges. The Seebeck coefficient (α) is precisely defined as the quotient of the electric potential difference (V) between two junctions and the temperature difference (T) between them.

$$\alpha = \frac{V}{\Delta T} \tag{1.1}$$

The Seebeck effect converts heat into electricity in a thermoelectric generator (TEG). The clean energy conversion rate is small, reliable, and smooth operation. Flexible wearable self-powered devices, space, and industrial waste heat recovery utilize thermoelectric generators.

The Seebeck effect generates electricity when different metals are heated. The temperature difference between the metal junctions determines the voltage of a TEG or Seebeck generator. Thermoelectric generators, which are solid-state heat engines, utilize p-type and n-type junctions. P-type elements exhibit positive Seebeck coefficients due to their doping with numerous positive charges or holes. N-type elements, which have a strong negative charge or electron doping, have lower Seebeck coefficients [5].

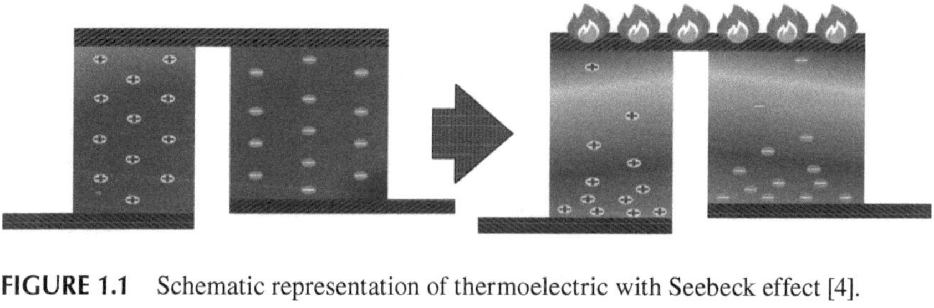

FIGURE 1.1 Schematic representation of thermoelectric with Seebeck effect [4].

1.3 FUNDAMENTAL PRINCIPLES

By thermoelectric phenomena, heat is transformed into electricity. Direct energy conversion is affected by the thermal, electric, Seebeck, and figure-of-merit coefficients of thermoelectric materials. These materials convert thermal into electrical energy. The main phenomena in thermoelectric devices are the Joule effect and thermoelectric effects (Seebeck, Peltier, Thomson) [6].

1.3.1 Seebeck Effect

TEG thermocouples are electrically connected in series and thermally in parallel due to the Seebeck effect. Because of their energy efficiency, easy maintenance, and extended lifespan, TEGs are used in many fields. Their size, power, and materials have been a topic of attention in energy harvesting for large and small applications in recent years.

Thermal generators use the Seebeck effect to generate energy. TEGs are chemical-free, simple to design, long-lasting, and low-maintenance. Figure 1.2a and b illustrate thermopiles with multiple series and parallel thermocouples (TCs). Two materials with different Seebeck coefficients from a thermocouple.

The Seebeck effect measures voltage and current. The equation calculates the thermoelectric generator's current density [7].

$$J = \sigma\left(-\nabla V + E_{emf}\right) \tag{1.2}$$

The temperature gradient T can be used to calculate electromotive field intensity using the material-specific Seebeck coefficient.

1.3.2 Peltier Effect

A thermocouple's Peltier effect emits heat at one junction and absorbs heat at the other when an electric current travels across it. The Peltier effect adjusts heat flow in thermal analysis and calorimetry. Peltier cooling devices are used to build superconducting magnets, isothermal microcalorimeters, and PC CPUs.

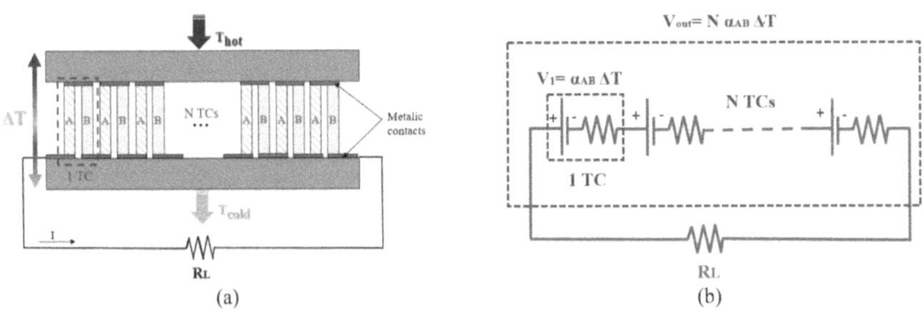

FIGURE 1.2 Standard thermoelectric generator: (a) Principle and (b) Electrical circuit [7].

A thermocouple's electromotive force (emf), as a function of metal electron concentration and temperature, is calculated using the non-equilibrium technique in current thermoelectricity theory. A thermocouple's emf is the difference between the absolute Seebeck coefficients (S) of two metals [8].

$$\text{Electromotive force } (\Delta U) = \int_{T_1}^{T_2} (\alpha_A - \alpha_B) dT \tag{1.3}$$

where

T_1 and T_2 = junction temperatures.

Each thermocouple arm has a unique temperature distribution, hence, U and α cannot be described by a universal temperature function. The Peltier and Seebeck effects are connected through the application of the novel thermodynamic model, which provides a quantitative explanation of the Seebeck coefficient. This produced the Peltier coefficient evaluation model, which affects temperature and Seebeck coefficient.

1.3.3 Thomson Effect

The non-linear Thomson effect is disregarded in most closed-form heat equation thermoelectric device models. A temperature-dependent Seebeck coefficient thermoelectric device generates electricity when charge carriers move with the voltage field while keeping a constant current. This non-linear factor's effect on thermoelectric performance can be neglected, resulting in inaccurate device expectations.

The Thomson coefficient is [8]

$$\tau = T.\frac{d\alpha(T)}{dT} \tag{1.4}$$

The Thomson effect enhances module power generation in thermoelectric materials, despite decades-old modeling assumptions. In addition to the Seebeck and Peltier effects, the Thomson effect requires reversible heating or cooling due to a conductor's current and temperature gradient.

P- and n-type semiconductors, top-side copper interconnectors, and two bottom-side interconnectors are arranged in each thermoelectric cooler's (TEC) thermoelectric pair. When an electric current moves from an n-type semiconductor to a p-type semiconductor, heat is absorbed on the cold side and transferred to the hot side, resulting in cooling. TEC transfers energy directly; is reliable, low-maintenance, and small; has no vibrating parts; and utilizes no refrigerants, unlike heat pumps and freezers. The device is adaptable since it is cooled by electricity.

A thermoelectric material's performance is measured by the dimensionless thermoelectric parameter ZT, where Z denotes its electric and thermal transport characteristics and T is the temperature (often room temperature) [9].

1.4 THERMOELECTRIC MATERIALS

Quantum wells, superlattices, wires, and dots manipulate electrons and phonons for the first time in low-dimensional materials. Quantum effects dominate, hence, changing the structural size can vary the electron and phonon energy spectrum, leading to new figures of merit (ZT) methods. Low-dimensional structures with the identical atomic structures as their parents are new in these techniques.

A theoretical and empirical study of a "new" thermoelectric material is possible for each size parameter. Thus, finding high ZT systems in low-dimensional architectures requires synthesizing and testing the thermoelectric characteristics of several bulk materials. Because their raw materials are simple and well-known, low-dimensional structures may be analyzed, predicted, and optimized. A transport mechanism, such as scattering phonons better than electrons, can be affected by classical size effects when quantum size effects are small [10].

Most small-scale and high-density cooling uses Bi_2Te_3-based thermoelectric materials due to their room-temperature capabilities. Low-temperature energy harvesting has also concentrated on Bi_2Te_3 materials. The fine-grained microstructures with (000l) orientation of the n- and p-type $Bi_2Te_{2.85}Se_{0.15}$ and $Bi_{0.4}Sb_{1.6}Te_3$ alloys improve thermoelectic and mechanical performance. The n-type alloy's thermoelecric properties differed from the p-type alloy, limiting its usage in thermoelectric devices. A comparison indicated that larger carrier concentrations lowered the thermoelectric properties of n-type hot-extruded materials. Cu dopant increased the power factor by decreasing carrier concentration, while excessive tellurium (Te) decreased carrier concentration but increased phonon thermal conductivity [11].

The remarkable capacity of these materials to immediately transform heat into electricity is notable. An oil-powered generator is 30% efficient. This is 70% of the chemical energy wasted as heat. Heat-to-power TE generators increase efficiency by 2.5%.

TE devices' low efficiency and contemporary materials' sustainability limit their adoption. To expand into non-niche applications, TE must compete with current technologies. Dimensionless TE figures of merit assess energy conversion efficiency. Biogenic Rankine and Kalina cycles convert heat with 40%–50% efficiency, hence, ZT must be at least 4. Costly, risky, rare-earth-based high-performance TE materials are a challenge.

Few 1D TE material reviews exist despite their availability [12]. Theories and thermodynamics have no ZT limits. High-performance thermoelectric materials and systems are needed as energy research accelerates. An optimal phonon-glass/electron-crystal material with a ZT of 4 has been achieved. These materials potentially solve many energy problems. A thorough search and careful investigation may yield these crucial elements for future TE devices. However, developing higher-efficiency thermoelectric materials and systems seems promising, with many materials still to be explored. Despite outstanding findings, thin films may be useful for small-scale electrical and optoelectronics applications with low heat and power generation.

High-performance bulk materials are needed for automobile and industrial power generation, as well as home freezer chilling. Theory-based band structure calculations and modeling are necessary to find feasible TE materials. Rapid but precise material characterization and result verification are also needed to advance this research. Higher-efficiency thermoelectric materials and technologies require multidisciplinary research. The process of creating "designer materials" for thermoelectrics may be applicable to other materials research [13,14].

1.4.1 Thermoelectric Materials Properties

The following are the properties of commonly used thermoelectric materials, and they can be considered based on the applications:

1. Thermoelectric figure of merit,
2. Thermoelectric quality factor,
3. Valley degeneracy and deformation potential,
4. Electrical conductivity and resistivity,
5. Mobility and charge carrier concentration, and
6. Mobility of a semiconductor.

The dimensionless ZT formula is used to evaluate thermoelectric materials for devices, aiming for the best efficiency (η_{max}) in converting heat into electricity. TE materials generate electricity from temperature gradients. Thermal gradients power thermoelectrics. The dimensionless figure of merit,

$$ZT = S^2\rho^{-1}\kappa^{-1}T \tag{1.5}$$

where

S = Seebeck coefficient,

ρ = Electrical resistivity, and

κ = Thermal conductivity.

A conductor's voltage is caused by its end temperatures. Seebeck coefficient is voltage-temperature ratio [15].

1.4.2 Thermoelectric Quality Factor

The quality factor (B) measures thermoelectric compound performance more holistically than zT.

$$\text{Compound quality factor}: B = \left(\frac{k_B}{e}\right)^2 \frac{T}{k_L}\sigma_{E_0} \tag{1.6}$$

where

σ_{E_0} and k_L = Transport coefficient (which is the carrier concentration-independent conductivity) and lattice thermal conductivity, respectively.

An ideal carrier concentration gives a compound's maximal potential zT. Using nanostructuring and flaw engineering to lower lattice thermal conductivity (k_L) can readily raise B.

1.5 VALLEY DEGENERACY AND DEFORMATION POTENTIAL

Improved quality can spur band structure engineering and the discovery of thermoelectric materials. Deformation potential theory may reveal that most good thermoelectric materials limit carrier mobility due to acoustic phonon scattering [16],

$$B \propto T\frac{N_v}{m_l^* \Xi^2} \times \frac{C_l}{k_L} \tag{1.7}$$

where

C_l = Elastic constants that will correlate with κ_L,

m_l^* = Inertial effective mass along conduction, and

Ξ = Elastic deformation potential coefficient that describes the electronic band energy change.

According to this formula, band structure engineers ought to focus on deformation and low effective masses.

1.6 ELECTRICAL CONDUCTIVITY AND RESISTIVITY

The n-type material $Bi_2Te_{2.7}Se_{0.3}$, after being hot-pressed, produced a figure of merit of 0.71 and a Seebeck coefficient of –185 µV/K (at 420K for 533K samples for 10 minutes). The maximum Seeback coefficient is –98 µV/K, electrical resistivity is –2.1 mΩ ·cm, and power factor is –0.4 W/mK^2 at ambient temperature [17,18].

1.7 MOBILITY AND CHARGE CARRIER CONCENTRATION

Trend in electrical conductivity (s) for $Bi_{1-2x}Li_xCa_xCuSeO$ and the new BiCuSeO sample. Temperature raises the electrical conductivity of the new BiCuSeO, indicating semiconductivity. Temperature decreases electrical conductivity in all co-doped samples, demonstrating that Li and Ca co-doped materials behave metallically over the examined temperature range. All co-doped samples have identical temperature-dependent electrical conductivity profiles. Higher electrical conductivity than pure BiCuSeO was seen in Li and Ca co-doped samples at all temperatures. Doping Li and Ca with $Bi_{0.84}Li_{0.08}Ca_{0.08}CuSeO$ doubles the electrical conductivity from 1.12 to 288 S/cm at 300 K.

The electrical conductivity value of $B_{i0.84}Li_{0.08}Ca_{0.08}CuSeO$ at 873 K is 96.97 S/cm, nine times that of pure BiCuSeO (11.3 S/cm). Li and Ca co-doping increases electrical conductivity at all temperatures. Hall experiments at room temperature explored the electrical transport improvements of Li and Ca co-doped BiCuSeO.

Carriers (n) and mobilities (m) in $Bi_{1-2x}Li_xCa_x$ CuSeO. Increasing the percentages of Li and Ca increases carrier concentrations. Li and Ca co-doped samples have higher carrier concentrations due to hole carriers from Ca2þ and Liþ doping at the Bi3þ site, leading to improved electrical conductivity in virgin BiCuSeO. Carrier mobilities are high in Li and Ca co-doped BiCuSeO samples. For instance, $Bi_{0.96}Li_{0.02}$–$Ca_{0.02}$CuSeO and $Bi_{0.92}Li_{0.04}Ca_{0.04}$CuSeO have the highest m values (8.06 and 7.51 cm^2/V s). Increased contaminant dispersion reduces carrier mobility compared to pure BiCuSeO following Li and Ca doping. Other ideally doped BiCuSeO should be compared.

Compared to heavy element doping, Li's atomic mass decreases carrier dispersion and enhances mobility. The CaO secondary phase drops m with Li and Ca doping fractions over 0.04, as detailed subsequently. Li doping's light atomic mass decreases carrier dispersion and enhances mobility due to Li-induced Fermi-level electronic states. The outer shell electrons of light element Li should push the Fermi level into the conduction band. Hall measurement shows Li–Ca co-doping in BiCuSeO changes carrier concentration and mobility, promoting electrical transport [19].

1.8 MOBILITY OF A SEMICONDUCTOR

Using a temperature difference, thermoelectric materials generate power. Total thermal conductivity ($\kappa_{e+}\kappa_l$) measures the sum of electronic and lattice conductivity. Additionally, the power factor ($PF=S^2\sigma$) is a crucial thermoelectric performance parameter. A large PF and low heat conductivity are needed for high ZT thermoelectric materials. 2D materials can change their ZT value to overcome this constraint.

Changing n- and p-type doping enhances thermoelectrics. The maximal room temperature ZT for n-type doping was 2.1 and for p-type was 2.4. ZT rises to 4.1 at 500 K with n-type doping. It was found that β-bismuthene's strong spin orbit coupling (SOC) significantly impacts its electronic band structure and transport coefficients. In n-type doped β-bismuthene, superior thermoelectric characteristics are observed compared to the p-type. It was discovered that α-bismuthene's thermoelectric characteristics at zero equilibrium can be analyzed using Green's function. The p-type doped material has greatly anisotropic thermoelectric properties, with a greater ZT value along the zigzag direction than the armchair directions [21].

Other thermoelectric material statistics and Barium Uranium Oxide (BaUO's) electrical resistivity s (5s, s: electrical conductivity) temperature dependency are shown in Figure 1.3. At low and high temperatures, Bi–Te and Si–Ge alloys are world-class thermoelectric materials. Semiconductive BaUO (Barium uranate) has a negative temperature coefficient of electrical resistance. $BaUO_3$ has four times the electrical resistance of thermoelectric materials. Figure 1.4 displays BaUO's Seebeck coefficient (S) with temperature, along with other thermoelectric materials. Electrons are the major charge carriers in BaUO because its Seebeck coefficient is negative at all temperatures. At 900 K, the absolute Seebeck coefficient peaks at 170 mV/K, which is a high value compared to modern thermoelectric materials.

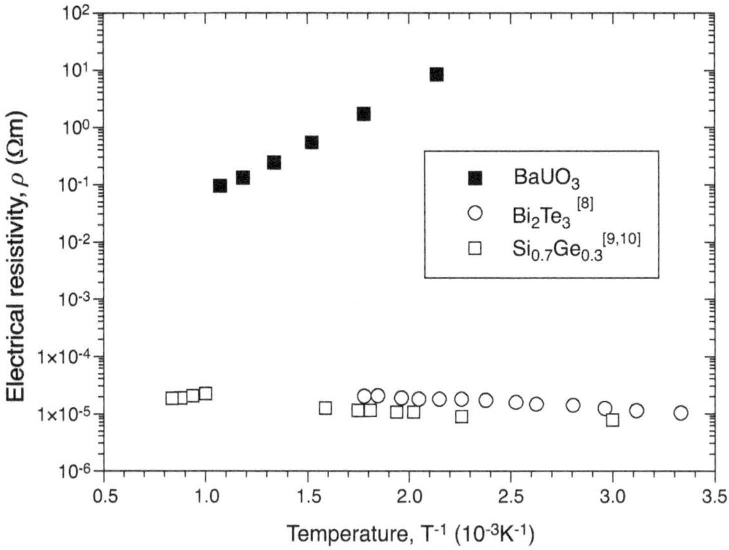

FIGURE 1.3 Electrical resistance (ρ) variation with temperature [22].

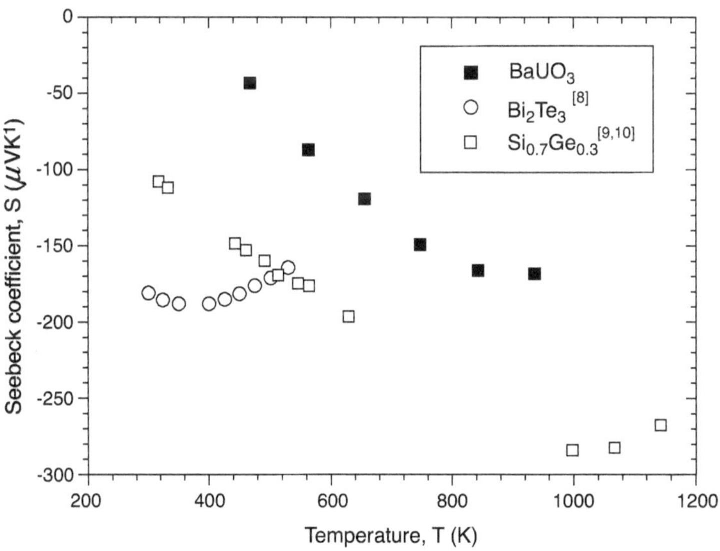

FIGURE 1.4 Seebeck coefficient (S) temperature dependence [22].

1.9 MECHANICAL PROPERTIES

Standard bulk thermoelectric devices include stiff electrode-bulk thermoelectric material connections. In practice, heat and external shear stress cause warping, fracturing, and peeling of the hard connecting electrode, increasing internal resistance. At compressive strength limits, the thermoelectric material splits, and gadget fails. All of these failure types diminish or reduce TEG output. Thin-film inorganic thermoelectric devices use wire or film electrodes like bulk devices. Bending thin-film thermoelectric materials require flexible substrates.

Service bending damages the film and peels off the bridge electrode connection, increasing internal resistance and lowering output. Organic materials are more flexible than thin films. It increases flexible thermoelectric device conversion efficiency and flexibility over bulk devices. As with bulk and flexible devices, bulk-leg flexible devices fail due to electrode material connection issues and flexible substrate fracture. TEG mechanical failure is caused by numerous variables.

Three categories define the main factors:

1. Materials' mechanical qualities matter for thermoelectrics;
2. Thermoelectric materials affect bulk; and
3. Flexible organic, inorganic thin-film, or bulk-legged devices' mechanical characteristics.

Electrodes create thermoelectric p-n junctions by bridging n- and p-type materials. Substrate mechanical properties affect TEG mechanical properties. Substrates make bulk-leg and thin-film inorganic devices flexible. Substrate fractures may rupture the thermoelectric material and separate it from the electrode, leading to TEG failure. A covalent semiconductor thermoelectric material is unplastic and brittle, making it rarely distort. Powder metallurgical or polycrystalline melting samples are thermoelectric. Pores, porous patches, agglomerates, big grains, and inclusions on the volume, surface, and edges reduce strength. TEG thermoelectric legs shatter from cross-dimensional hybrid flaw.

Thermal materials, ceramic plates, electrodes, and barrier layers make up bulk thermoelectric devices (Figure 1.5a). Thermal stress damages thermoelectric leg-electrode connections. Cracking is likely near the connecting layer's corner due to significant shear stress. Figure 1.5b reveals that the hot side's area (with the higher Cu connection) and lower surfaces have the maximum thermal stress. High stress in these area strains and cracks the ceramics-copper electrode connection, affecting thermal conductivity. Electrical conductivity diminishes due to microcracks in regions under stress. By raising leg height, using asymmetric legs, etc., geometric optimization (Figure 1.5c) can reduce thermal stress on the thermoelectric leg edge. Thermal stress varies with position because thermoelectric legs have axial temperature gradients.

Different thermoelectric leg locations and forms have different maximum thermal stresses. Warpage and thermal stress distribution cracks during service increase contact electrical and thermal resistances, reducing output. In Figure 1.5d, thermoelectric materials have joint shear strengths ≤ 60 MPa. Mg_2Si/Cu, $CoSb_3$/Ti/Mo–Cu lost shear strength with age. After 16 days of ageing with $Al_{66}Si_{34}$ as GeTe's barrier layer, joint shear strength increased from –26.6 to –41.7 MPa. The thickness of the aged diffusion layer and the contact resistance both grew. There were no microcracks at the interface. Interface microcracks generate most of the joint shear strength loss.

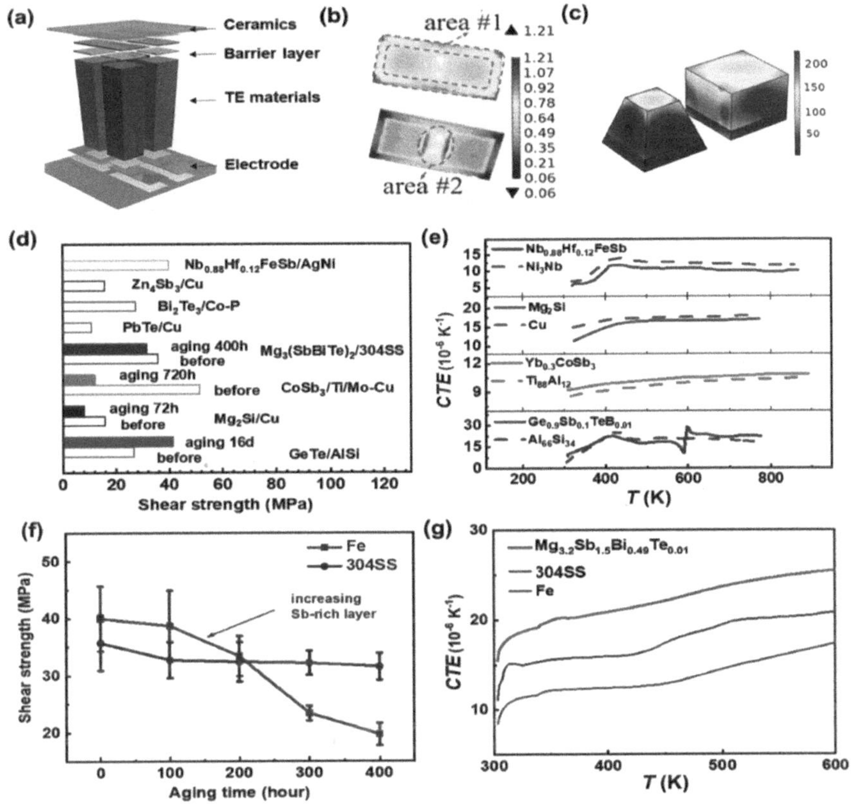

FIGURE 1.5 (a) TE device construction diagram, (b) distribution of TEG thermal stress, (c) TEG thermal stress distribution with segmented asymmetry, (d) joint shear strength, (e) thermal expansion coefficient, (f) Mg_3(SbBiTe)2/304SS and Mg_3/Fe shear strength, and (g) thermal expansion coefficient [5].

Thermal stress-induced crack formation and extension are minimized by electrodes and barrier materials featuring thermoelectric thermal expansion coefficients (Figure 1.5e). $Mg_{3.2}Sb_{1.5}Bi_{0.49}Te_{0.1}$ has Fe and 304SS (304 Stainless Steel) barrier layers (Figure 1.5f and g). Fe barrier layer shear strength depreciation exceeds 304SS after 400 h at 523 K [5].

1.10 ELECTRICAL PROPERTIES OF THERMOELECTRIC MATERIALS

To improve ZT values, increase the power factor ($S^2\sigma$) and decrease heat conductivity. Most research currently focuses on reducing the thermal conductivity of ZnO-based TE materials. Optimizing ZT values requires power factor enhancement. Despite having a high Seebeck coefficient (~340 V/K), ZnO ceramics' electrical conductivity (~10 S/m) is insufficient for practical usage.

Expand donor dopant solubility to enhance the power factor. Carrier concentration augmentation must be carefully designed due to the Seebeck coefficient-electrical conductivity tradeoff. Electrical conductivity increases with high carrier concentration and reduces the Seebeck coefficient. Normal carrier concentration maximizes the power factor. In addition to carrier concentration, Hall mobility increases electrical conductivity. Low Hall mobility in ZnO ceramics is caused by the grain boundary Schottky barrier, which hampers electron transport. Thus, increasing ZnO ceramic electrical conductivity requires lowering the grain boundary Schottky barrier.

The temperature dependency of σ for Spark Plasma Sintering (SPS), NC (sample name), and A (sample name) samples is shown in Figure 1.6a. Increasing temperatures decreased σ in SPS and NC samples, showing metal-like behavior. The SPS and NC samples have σ values of 2.3×105 and 1.9×10^5 S/m at ambient temperature, respectively, three orders greater than the sample. The SPS sample had a greater σ than the NC sample at all temperatures, as seen in the inset. Sample A had the lowest conductivity (1.6×10^2 S/m) and rose with temperature, indicating a thermally activated semiconductor behavior.

The maximum σ in the SPS sample was determined by assessing carrier concentration (n) and Hall mobility (μ) (Figure 1.6b). Throughout the recorded temperature range, the n values for SPS and NC were practically independent. The NC sample has $n=1.6\times10^{20}\,cm^{-3}$, while the SPS sample has $2.6\times10^{20}\,cm^{-3}$, which is 63% higher. In both SPS and NC samples, μ remained constant at low temperatures but decreased at high temperatures due to lattice vibration. The SPS and NC samples had μ values of 63 and 74 $cm^2/V{\cdot}s$ at ambient temperature. Low carrier concentration prohibited Hall measurement equipment from detecting sample A's electron density. Carrier concentration, not Hall mobility, made SPS conductivity higher than NC.

In Figure 1.7a, all samples displayed negative Seebeck values across the temperature range, demonstrating n-type conduction. SPS, NC, and A samples have Seebeck coefficients of around 144 μV/K at ambient temperature. Due to carrier

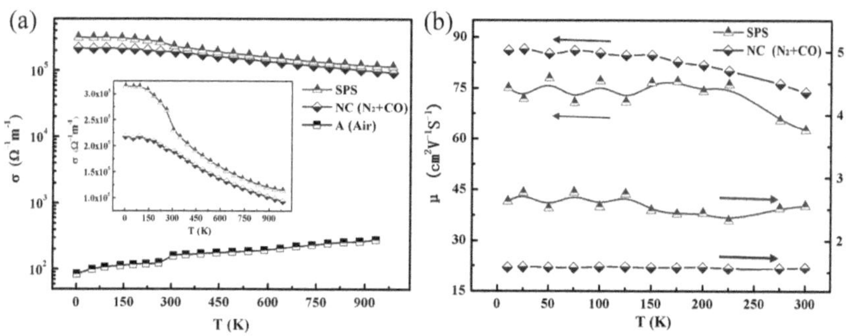

FIGURE 1.6 (a) Temperature-dependent electrical conductivity of SPS, NC, and A samples; SPS and NC samples' electrical conductivity is inset. (b) The SPS and NC samples' Hall measurements [20].

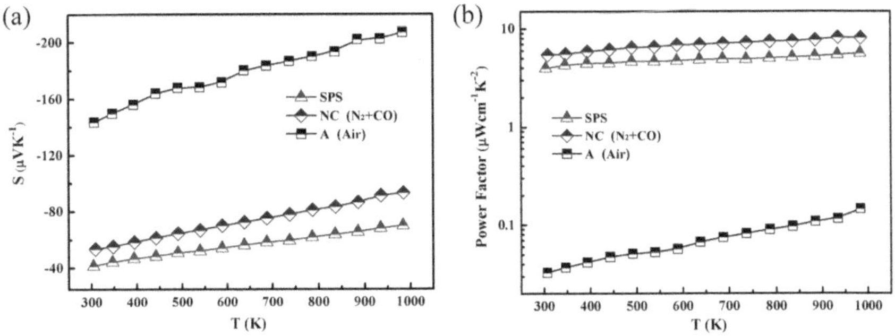

FIGURE 1.7 (a) The Seebeck coefficient and (b) power factor of the samples SPS, NC, and A [20].

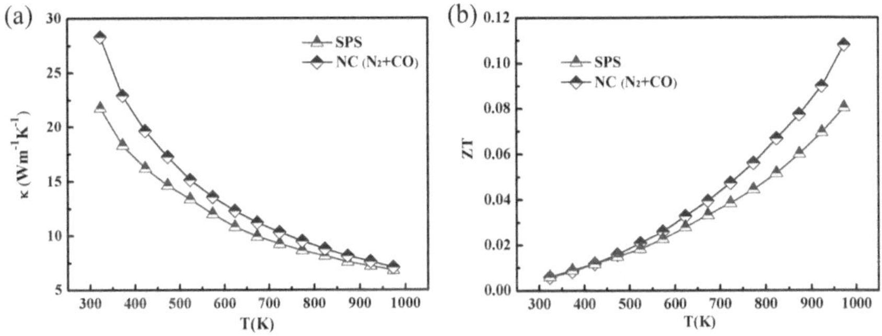

FIGURE 1.8 (a) The thermal conductivity and (b) ZT values of the samples SPS and NC [20].

concentration, SPS has the lowest Seebeck coefficient (S). The Seebeck coefficient inversely affects carrier concentration. Thus, the sample SPS Seebeck coefficient was the lowest at the highest carrier concentration.

The temperature-dependent power factor, $S^2\sigma$, is shown in Figure 1.7b for the three samples. Sample A has lower power factors than SPS and NC due to electrical conductivity.

Figure 1.8a shows the thermal conductivity (κ) of SPS and NC samples. SPS is less thermally conductive than NC. Due to its lower power factor, the sample SPS's dimensionless figure of merit, ZT, was lower than the sample NC's (Figure 1.8b) [20].

1.11 ADVANCED THERMOELECTRIC MATERIALS

A variety of solid compounds can convert heat and electricity as thermoelectric materials. This feature creates temperature gradients for heating and cooling equipment and generates electricity from waste heat. The thermopower Seebeck coefficient (μV/K) measures voltage generation per temperature change. High figure-of-merit thermoelectrics have strong electrical conductivity, thermopower, and low thermal conductivity.

When temperature gradients cause carriers to diffuse to the cold side, the module generates thermoelectric voltage. ZT has remained around 1 for prototype materials at all temperatures for decades. Antimony and bismuth tellurides are utilized at 300 K, lead and skutterudite antimonides at 650 K, and silicon-germanium alloys at 1000 K.

Decoupling the interrelated S, σ, and κ is a challenge in developing better thermoelectric materials. Nanostructuring reduces lattice heat conductivity. Numerous nanoscale grain boundaries disperse heat-carrying phonons. This method considerably boosted Bi–Sb–Te alloy ZT. A 1.5-ZT bulk sample was generated from the powder after high-temperature vacuum pressing. The PbTe–AgSbTe pseudobinary system possesses coherent nanoscale inclusion phases with low thermal conductivities and high ZT (>1). Finally, thermoelectric materials with customized electrical topologies created 1.5-ZT thallium-substituted PbTe.

Bulk thermoelectric materials are made using powder metallurgy and high-temperature melts. To maintain physical properties, high-purity elemental precursors (Table 1.1) are needed to lower conductivity and charge carrier concentration. For purposeful doping/substitution and electrical band structure tailoring investigations, high-purity starting materials provide precise composition control.

Polyol reduction method and solution-phase solvothermal synthesis are other thermoelectric material processes. However, solvent, surfactant, and reducing

TABLE 1.1
Thermoelectric Materials Melting Point and Their Properties [23–29]

S. No.	Name of the Material	Melting Point (°C)	Properties
1	Antimony(III) selenide	611	This chemical contains antimony oxidation state +3 and selenium −2. The black color and semiconducting characteristics of this chemical and similar compounds indicate covalent bonding.
2	Antimony(III) telluridepowder	620	As a topological insulator and narrow-gap semiconductor with a band gap of 0.21 eV, Sb_2Te_3 has thickness-dependent physical characteristics.
	Bismuth selenidegranular (melted)	710	Because selenium vacancies contribute electrons, stoichiometric bismuth selenide should be n-type and have a 0.3 eV gap.
	Lead(II) selenide	1078	It produces cubic NaCl crystals and has a 0.27 eV direct bandgap at ambient temperature.
	Lead(II) telluride	924	It crystallizes in NaCl with Pb atoms as cations and Te as anionic lattices.
	Silver(I) telluride	955	The semiconductor silver telluride can be doped n- and p-type. Stoichiometric Ag_2Te conducts n-type. Silver is lost when heated.

conditions vary greatly when precipitating discrete granules from cationic solutions of bismuth nitrate and lead acetate. Using phonon scattering, chemical nanoscale thermoelectric materials reduce the thermal component and relate electrical and thermal conductivities. Waste heat can be converted into power by thermoelectric materials, improving engine efficiency. This approach can also scavenge energy from electronics and automobile emissions.

1.12 FABRICATION AND CHARACTERIZATION TECHNIQUES

Construction of microscale thermoelectric devices involves air bridges. A photoresist sacrificial layer has two fabrication difficulties. The first challenge after patterning is the Au top electrode bridge profile thickness homogeneity. The shadowing effect thins the Au electrodes on the sacrificial layer sidewall, reducing the sharp edge connection. Manufacturing can fracture the Au electrode and increase device internal resistance. During the following development process, the sacrificial layer may be removed. Good fabrication fixes these issues. This method is dangerous but effective. Thin metal electrode resistance increases with a rough plasma asher-etched surface underexposure and poor exposure depth.

Other techniques increase fabrication time and expense. The next development step may eliminate the untreated sacrificial layer. A good fabrication technique solved these difficulties. The challenging exposure depth and rough plasma asher-etched surface underexposure increase the thin metal electrode's resistance. Additional steps raise fabrication costs and time [30].

To make thermoelectric materials, the following methods are used:

1. Non-contact,
2. Inkjet printing, and
3. Photoresist melting.

1.12.1 Non-contact Technique

Although inkjet or aerosol printing has lower throughput than contact printing, it requires less material, has a lower contamination risk, offers pattern flexibility for prototyping, and provides higher resolution. Maskless methods simplify substrate-based digital design transfer without additional steps.

1.12.2 Inkjet Printing

Inkjet printing sprays ink droplets from print head orifices onto the substrate instead of touching it. This low-cost, high-resolution manufacturing technology deposits liquids over large areas with significant versatility and personalization. Inkjet printing can be continuous or drop-on-demand.

From the reservoir to thin nozzles, continuous inkjet (CIJ) creates a liquid filament (Figure 1.9a and b). Ultrasonic waves from piezoelectric material deformation split the filament into homogeneous droplets. Piezoelectric droplet production

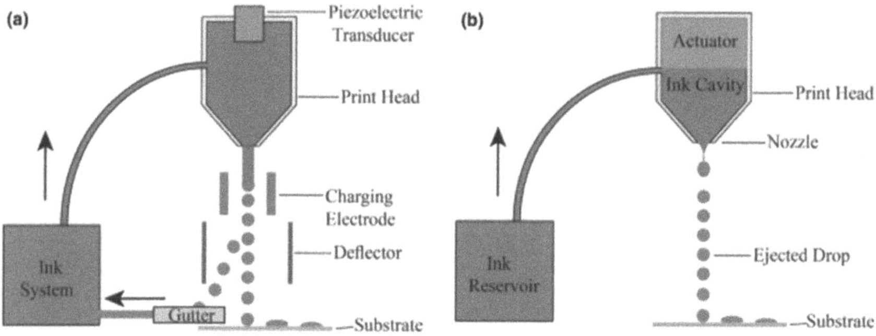

FIGURE 1.9 Schematics of continuous (a) and drop-on-demand (b) inkjet printing [31].

modifies droplet dimensions. After that, digital data charges ink droplets between plates. Electrostatic fields between high-voltage deflector plates deflect charged droplets towards the substrate. Since orifice occlusion is unlikely, continuous flow works for volatile solvents. Extended recycling might taint the ink, stick to plates, and damage the printer [31].

1.12.3 Photoresist Melting

Micro-TEDs have substrate-perpendicular transverse heat flow. Four-inch SiO_2/Si wafers with six micro-TED modules and 10,082 P–N thermocouples are shown in Figure 1.10. UV lithography, magnetron sputtering, photoresist melting, and five steps make up the bottom-up method (Figure 1.10a–e).

Shorter thermoelectric legs (L) enhance power density or heat pumping without electrical and thermal contact resistance. Interfaces had 60 nm Cr bonding layers (Figure 1.10g). Each step has a 100 nm Au layer on the thermoelectric material to avoid oxidation and minimize contact resistance. Device construction prioritizes top electrode connection reliability. The 4 µm-thick sacrificial photoresist was shaped and smoothed using UV lithography and photoresist melting to support the top electrode [32].

1.13 Prospects of Thermoelectric Materials for Clean Energy Harvesting

Waste Heat Recovery (WHR) practicality depends on the waste heat (WH) stream quality, volume, and condition (gas, liquid, solid) [29]. WHR uses heat-to-heat, work, and power. Figure 1.11 shows three WHR techniques for traditional power systems. Heat is transferred from WH to another process stream using typical heat transfer methods in the first instance, such as fuel–air mixture, fractionation column oil feed, melting boiler load, and boiler feed water.

Many power plants transform turbine work into electricity using classical power generation or heat-to-work with an electric generator. According to Figure 1.11, three stacked cycles transform energy into a higher-quality form. Losses from energy quality improvements lower energy performance and conversion.

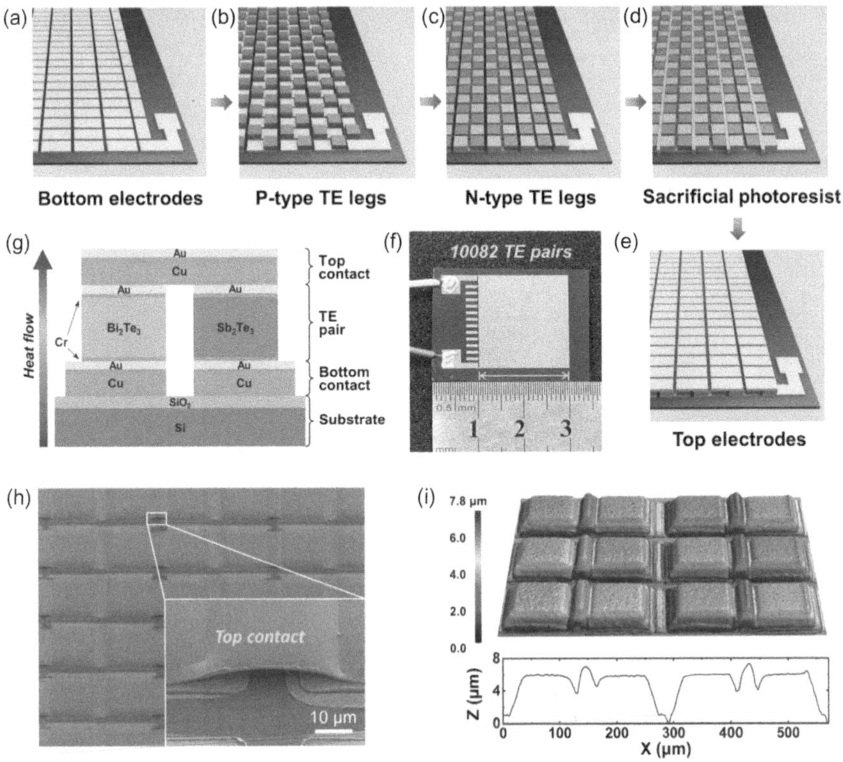

FIGURE 1.10 Micro-TED fabrication and characterization. (a) Patterned SiO_2/Si bottom electrode. (b) P-type thermoelectric legs patterned. (c) N-type thermoelectric legs patterned. (d) Construction of support adhesive. (e) Top-patterned electrodes. (f) The micro-TED module has 10082 P-N thermoelectric couples. (g) Multilayer material arrangement of a single P–N thermoelectric pair and side profile. (h) The P–N thermoelectric pairs configuration and air bridge top electrode SEM picture. (i) Laser micrograph of micro-TED thickness. Stretching the image's Z-axis [32].

TEG maximizes efficiency and provides most of the world's energy demands without carbon. TEGs cool one side of a semiconductor material and heat the other. As shown in Figure 1.12, temperature gradients increase electrical potential.

1.14 RADIO WAVE ENERGY HARVESTING

Radio waves can be captured anywhere by using our iPhones to search, even on planes and trains. Most radio waves are electronically generated and transmitted in a specific frequency band. Commercial wireless power supply devices broadcast energy-rich radio waves. Radio wave energy harvesters require different circuits than vibrational energy harvesters due to frequency differences. For instance, an energy storage increasing circuit must address power loss. Energy harvesting from high-frequency bands reduces antenna size for 5G, but circuit

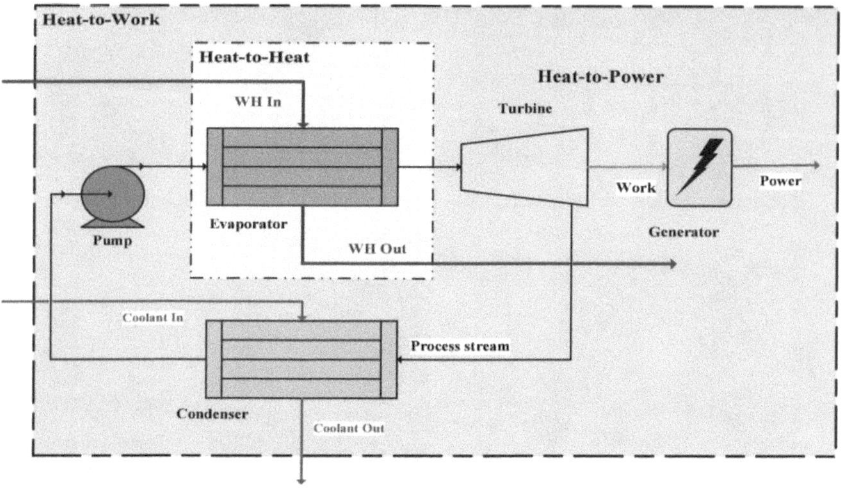

FIGURE 1.11 Waste heat recovery methods [33].

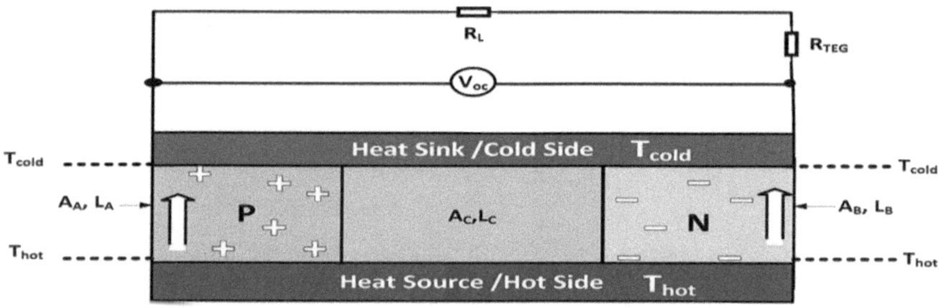

FIGURE 1.12 Schematic diagram of thermoelectric generator (TEG) [33].

design must balance application needs. Rectifier circuitry converts radio waves into DC current in receiver antennas. From device technology to evaluation, radio wave energy harvesting antennas have been studied.

GaAs Schottky barrier diodes are replaced with a high-sensitivity reverse diode that uses III–V semiconductor nanowires to transform weak radio wave energy into electricity. P–N junction body-tie SOI-FETs with steep current characteristics exhibit superior rectification and can collect radio wave energy. Flexible 2D semiconductor molybdenum (IV) sulfide (MoS_2) diodes have shown promise for 2.4 GHz rectennas.

1.15 FEATURES OF THERMOELECTRIC ENERGY HARVESTING

Multiple surveys show that we waste 70% of fundamental energy as waste heat. Surveys indicate heat dissipation below 100°C. Thermal energy harvesting, which transforms heat into electricity, is becoming popular for IoT devices. Conductor ends generate a voltage ΔV proportional to the induced temperature differential ΔT. In a 1°C per unit length temperature differential, the power factor (PF) ($S^2\sigma$ in the numerator) represents power output (W) [34].

1.16 ENVIRONMENTAL BENEFITS OF THERMOELECTRIC ENERGY HARVESTING

Energy (especially fossil fuels) has increased carbon dioxide emissions and environmental challenges that affect every country on Earth. However, wasted heat can be used for energy. Manufacturing machine production waste heat is converted into power using thermoelectric technology, reducing fossil fuel use as an alternative energy source. Continuous heat release makes compact, flexible, high-absorption thermoelectric systems excellent for machines. Air conditioners always produce too much "waste heat"—low-temperature heat. A thermoelectric generator may generate electricity from waste heat. Practically, one-third of fuels supply energy, and two-thirds are waste heat. The environment loses 20%–50% of energy as waste heat. Operating electrical machines emit waste heat. More heat is wasted than used for production, reducing electricity efficiency. Air-conditioning systems release building heat into the environment, which largely harms climate change. The "Heat Island Effect" can result from installing heat traps [35].

Global warming is caused by waste heat released into the environment. Electric gadgets, both big and small, waste heat. A motorcycle's exhaust system generates heat that can charge phones and GPS devices. Low-maintenance thermoelectric generators are free and pollution-free. Most government-sponsored thermoelectric research in the past 10–15 years has concentrated on power generation. Most of this research focuses on energy efficiency. The automobile exhaust and radiator lose most of their energy as heat.

Lost heat is converted into energy using thermoelectric waste heat recovery. The commercial and military use of energy harvesting, along with its capacity to use components continuously, off-grid, and for long periods, has garnered attention. Wearable electronics also seek harvesting devices to power radio communications, phones, mobile computers, and more. Night-day temperature fluctuations may one day power outdoor applications. Building a long-lasting device in harsh conditions is challenging. Their designs must also take advantage of environmental energy.

Thermoelectric energy is intriguing and promising for both large and small applications. Thermoelectric generators can capture heat from several sources [36,37].

1.17 ENVIRONMENTAL ISSUES

Transitioning requires tremendous effort and groundbreaking new technology. Hydraulic, nuclear, geothermal, wind, solar, biofuels, fossil-fuel power generation with carbon dioxide capture and sequestration, and conservation measures are carbon-free because the technology is accessible. This UN-commissioned

World Energy Assessment Overview details the worldwide energy portfolio and costs with current data. Hydropower accounts for 2% of global energy (0.3 TW).

The world has 10 GW of geothermal electricity. By 2007, wind power had 100 GW of worldwide generating capacity and an unsustainable 30% annual capacity increase. In locations with high average wind speeds, wind power is inexpensive but intermittent and can only supply 20% of electric power demand while maintaining system reliability. In 2007, solar power generated just under 10 GW worldwide, but new capacity is growing rapidly at 30% per year.

The classic TE materials Bi_2Te_2, PbTe, and Si–Ge alloys were developed in the 1950s and 1960s. Their performance (as measured by the figure of merit zT) was improved to values close to unity, and the first practical applications of thermoelectricity were demonstrated. However, the field stagnated with no significant improvements in TE performance or new insights into the TE conversion process.

Environmental problems include pollution, hazards, and ecological degradation. Society has changed due to technology, and environmental issues now significantly influence both humanity and nature. The risk and reality of environmental degradation have escalated over the past two decades.

The environmental impact of human activities has increased due to factors such as population growth, consumption, industrial activity, and others. Micro or hazardous air pollutants, toxic chemical substances that are harmful in small doses, and globally significant pollutants like CO_2, the main greenhouse gas responsible for global warming, have joined conventional effluent gas pollutants like SO_2, NO_2, particulates, and CO.

As environmental science and industrial processes and structures advance, new environmental issues arise, such as in the energy sector, where road transport of industrial goods and individual car travel have increased road traffic and NO_x and volatile organic compound emissions. This can be addressed using thermoelectric generators in energy generation [38].

Two causes conspired to enhance TE research in the mid-1990s. First, societies became increasingly concerned with alternative energy and energy conservation as fossil fuels became scarce and caused significant environmental damage. Second, scientific and technological advances, such as a better understanding of mesoscopic transport processes, the central role of nanoparticles as structural elements, advanced synthesis techniques, and new powerful analytical tools to understand material structure, were crucial. Once these two factors combined, funding organizations increased support for TE research, sparking a new generation of enthusiastic researchers worldwide.

Thermoelectricity, with better TE materials, could have a significant impact with breakthroughs in transport phenomena, efficient simulation tools, and novel synthesis methods. Since the mid-1990s, thermoelectricity has been reinvigorated, leading to new families of efficient TE materials and advances in classical TE materials.

These rapid research discoveries revealed new physical mechanisms that impacted the development of new TE materials, many of which have yet to be critically reviewed. The global effort has increased the performance of TE materials, offering new bulk TE applications. More energy efficiency is still needed [39].

The global challenge of building a sustainable energy generation, storage, and distribution infrastructure requires massive transnational investments in energy technology research and development to provide sufficient energy on time and at scale, with minimal environmental impact and economic and societal disruption.

Define the energy problem's current condition, future needs, and temporal constraints, then define the technical development speed. Global primary power consumption is approximately 14 terawatts (TW, 14×10^{12} J/s), with fossil fuels (oil, coal, natural gas) producing over 80%, and renewable sources (geothermal, wind, solar, biofuels) contributing less than 1%.

As China and India industrialize, global power demand is expected to double to 25–30 TW by 2050 under the "business as usual" development scenario. If fossil fuels continue to generate ~80% of power, carbon dioxide emissions will pose a severe environmental threat to civilization.

The Intergovernmental Panel on Climate Change (IPCC) evaluated the environmental implications of potential CO_2 emission trajectories from carbon-based energy sources. The 2007 results assessment report is the latest. Despite no carbon dioxide tipping point, these papers reveal that agriculture, water resources, and human health will suffer substantial and likely irreversible environmental damage if the world average temperature climbs beyond 2.5°C–3.0°C.

Recent studies anticipate this temperature spike will occur at 350 ppm of average atmospheric CO_2, whereas a more conservative EU study estimates that 550 ppm is needed. These forecasts are close, and stability or reversal requires a rapid global shift to renewable energy sources, with 20 TW of low-cost, carbon-free power by 2050.

Carbon dioxide capture and sequestration (CCS) in fossil-fuel power generation could cut emissions without increasing energy infrastructure. Practical and economic concerns with CCS are major. Carbon dioxide capture technology would triple power prices, and geological sequestration is unproven.

The parasitic energy load of CO_2 capture equipment requires 30% additional fossil fuel. To eliminate transport sector carbon dioxide emissions, carbon-dioxide-neutral cellulosic biofuels, synthetic hydrocarbon fuels based on atmospheric or captured CO_2 and water (e.g., solar photocatalysis), hydrogen-driven vehicles and infrastructure, and/or electric vehicles with carbon-free power are needed.

Energy corporations working with governments and global collaborations face scientific, technological, political, and economic challenges in carbon dioxide capture and sequestration [40].

Building a sustainable energy system is hard. TE R&D—especially when using waste heat—is partially driven by the notion that it can help attain this goal. Recent analysis reveals that their contribution is small, even if huge gains in the figure of merit, ZT, etc., are projected; mechanical engine efficiency advantages will persist.

In comparison to the reference situation, the most sustainable technology or resource application has the most benefits, the fewest negatives, and a favorable grade. TEG may use waste heat in various systems, comparing three TE applications with distinct beneficiaries. Bi_2Te_3 for TE and CdTe for PV use tellurium, which can also be used in other energy technologies.

TE's ability to make sectors and the energy system more efficient and sustainable can be assessed by TE application decrease rates and sector energy output or consumption. Additionally, thermal condensation plant power generation, as well as industrial energy use and recovery, are estimated.

Low-CO_2 energy systems require small- and medium-scale CHP employing liquid or gaseous fuels. Internal combustion engine transport and woodstoves can be reduced. Using biomass more efficiently saves the most. Low temperatures and energy densities characterize large waste-heat potentials, such as condensed steam and massive thermal condensation power plant exhaust gases.

To optimize TE materials, modules, and systems, and to find ecologically and sustainably optimal applications, information on additional environmental impacts, sustainability (costs, employment, working conditions, and technological accessibility), and competing technologies is needed. Integrating sustainability assessments into TE R&D will efficiently obtain this knowledge [41].

1.18 SUSTAINABLE PRODUCTION TECHNIQUES

With improvements in wireless technology and the IoT, researchers have developed wireless sensors for many uses. Wireless Sensor Network (WSN) devices for military, home appliances, industrial applications, and wearable health monitors are examples. WSN monitors temperature, pressure, steam, bridge structural health, and more, and is increasing rapidly. Condition sensors need constant, battery-free electricity to monitor data.

WSN battery maintenance might be difficult in remote, inaccessible locations. A self-sufficient power supply for WSNs is needed to eliminate device replacement/recharging and enable continuous physical data monitoring. TEGs convert heat to electricity without moving parts and are eco-friendly. TEGs are easy to install and use waste heat. TEGs may provide a useful, self-sufficient power supply for remote WSN devices.

The prototype radial TEG device offers several benefits:

1. Stencil printing and scaling,
2. High-aspect-ratio thermoelectric element packing, and
3. No high-temperature (500°C) or extensive processing.

Reduce energy utilization and enable large-scale TEG production with a low thermal budget manufacturing technique. Radial TEGs on flexible Kevlar can capture heat from pipes. Energy efficiency promotes all energy sectors and sustainability, as waste heat or improved industrial heating replaces fossil fuels. Printable, scalable TEG devices may power industrial condition monitoring sensors safely and sustainably, instead of using stiff, flammable batteries [42].

Due to economic and technical constraints, TE technologies are undermarketed. TE technology research has focused on improving electrical transport and figure of merit, rather than material processing or manufacturing costs. TE material and device Conventional Manufacturing (CM) is multi-stage, difficult, and labor-, time-, and energy-intensive. Manufacturers' difficulties limit the industrial deployment of TE technologies. Engineering design thinking and production challenges have shifted due to modern Additive Manufacturing (AM) technologies. AM can improve thermoelectric technology's CM deficiencies with its many capabilities.

AM has sustainability, technical, and economic benefits. AM implements circular economy and sustainable production practices. AM design freedom improves product and process efficiency and adaptability. Design freedom allows for lightweight, multipurpose structures with fewer parts and easier installation.

AM uses fewer raw materials and reduces waste, making it a resource-efficient production method. AM could also support on-demand and distributed production. Customer-on-demand production minimizes AM supply chain costs and transportation-related CO_2 emissions.

AM is resource-efficient because its additive nature maximizes raw material usage and minimizes waste. AM may also influence on-demand and distributed manufacturing. By producing near customers, AM may reduce supply chain expenses and CO_2 emissions [43].

Multiobjective, transdisciplinary manufacturing sustainability evaluation is difficult. Sustainability is hard to measure. Although energy usage decreases, the sustainability of industrial processes/sectors cannot be directly compared.

Chemicals and petrochemicals, iron and steel, nonmetallic minerals, and pulp and paper are the largest industrial energy users. Industrial subsectors need greater energy efficiency. Usable energy generation to energy input is energy efficiency. Manufacturing innovations improve industrial energy efficiency.

Renewable energy sources can reduce "energy consumption", which can increase energy efficiency, economics, and reduce greenhouse gas emissions, but they do not improve end-use process efficiency. Efficiency and renewables lower energy costs and environmental impact. Renewables cost more per unit than energy efficiency. Optimization of energy efficiency can precede the adoption of renewables. Any energy conversion method is limited by thermodynamics, and real-world technology cannot reach that limit. Energy efficiency remains low.

Manufacturing has improved energy efficiency for decades, but less has been done to recover waste energy. Mechanical energy, such as vibration and motion, sound, light, and thermal energy, such as heat and temperature gradients, are released into the environment as waste during manufacturing. Manufacturing generates a significant amount of waste energy.

Small waste energy recoveries are economically and environmentally beneficial. Harvesting waste energy reduces energy costs and improves energy efficiency for sustainable manufacturing. Energy from manufacturing waste might be considered "renewable" or energy efficiency-boosting. Along with manufacturing process optimization, waste energy collection devices can enhance energy efficiency in sustainable manufacturing processes.

Smart manufacturing needs energy harvesting. By enhancing intelligence, these technologies could boost large-scale industrial efficiency. Energy harvesting reduces greenhouse gas emissions beyond just energy recovery. Environmentally friendly manufacturing can harvest waste energy from manufacturing processes, convert it into electricity, process it in power conversion devices, and store it in a battery for later or immediate use in IoT systems.

In thermoelectric energy harvesting, a large temperature gradient provides a persistent voltage, necessitating massive heat flows. Pyroelectric energy harvesting produces a rare transient voltage when temperature changes occur. Temperature changes with distance and time in both thermoelectric and pyroelectric effects. As temperature stabilizes, the pyroelectric voltage drops.

1.19 THERMAL ENERGY HARVESTING IN SUSTAINABLE MANUFACTURING

Iron and steel, chemicals, and petrochemicals generate much of the waste heat. Waste heat is carried by exhaust gas, cooling air/water, hot oil, hot steel, etc. Industry uses 25.6% of EU energy. Chemicals and steel use the most, followed by paper, nonmetallic minerals, and food. These five sectors consume two-thirds of industrial energy. Industries demand varying levels of heat.

Food, tobacco, textiles, and leather require 100°C, pulp and paper require 500°C, and metals, iron, steel, and nonmetallic minerals require temperatures beyond 500°C. Heat from exhaust gases, cooling air/liquid, etc., accounts for 20%–50% of industrial energy use, according to the SPIRE Roadmap (2020). Figure 1.13 contrasts the US business sector's usable energy and energy losses, highlighting waste heat energy potential.

All manufacturing generates waste heat. Manufacturing with waste heat energy saves energy for sustainability. Harvesting thermoelectric energy is difficult due to its poor conversion efficiency of 4%–6%, with a maximum of up to 10%. Further R&D is needed to improve thermoelectric energy conversion efficiency and cost-effectiveness for sustainable waste heat recovery with a payback period of less than 3 years.

Industrial energy use accounts for one-third of global energy consumption. In 2016, over 72% of worldwide electrical energy generation came from industrial waste heat. Efficiency is considerably reduced by manufacturing heat losses. Sustainable manufacturing requires energy reduction, which starts with improving production efficiency. Thermodynamics limits energy efficiency in practice. Achieving energy efficiency in sustainable production requires waste energy harvesting.

Even a small amount of waste energy recovery improves environmental and economic sustainability. Manufacturing generates waste heat and mechanical vibration. Heat energy can be converted to electricity by thermoelectric, pyroelectric, thermomagnetic, piezoelectric, electromagnetic, electrostatic, and other methods. Waste energy harvesting technology and applications should improve in the next 5–10 years [44].

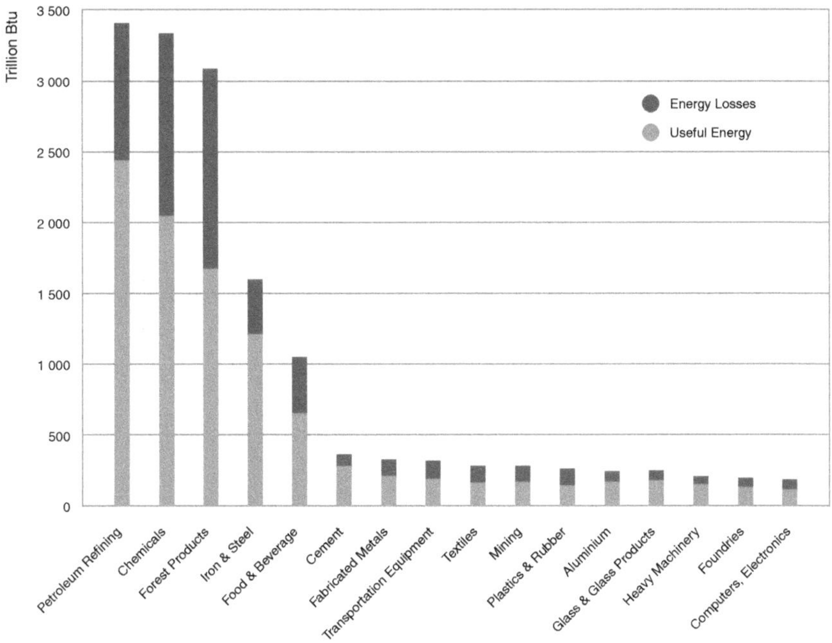

FIGURE 1.13 Overview of energy losses within sectors [44].

REFERENCES

1. Zulkepli, N., Yunas, J., Mohamed, M. A., & Hamzah, A. A. (2021). Review of thermoelectric generators at low operating temperatures: working principles and materials. *Micromachines*, *12*(7), 734.
2. Jouhara, H., Żabnieńska-Góra, A., Khordehgah, N., Doraghi, Q., Ahmad, L., Norman, L., ... & Dai, S. (2021). Thermoelectric generator (TEG) technologies and applications. *International Journal of Thermofluids*, *9*, 100063.
3. Champier, D. (2017). Thermoelectric generators: a review of applications. *Energy Conversion and Management*, *140*, 167–181.
4. Tohidi, F., Holagh, S. G., & Chitsaz, A. (2022). Thermoelectric generators: a comprehensive review of characteristics and applications. *Applied Thermal Engineering*, *201*, 117793.
5. Bao, X., Hou, S., Wu, Z., Wang, X., Yin, L., Liu, Y., & Zhang, Q. (2023). Mechanical properties of thermoelectric generators. *Journal of Materials Science & Technology*, *148*, 64–74.
6. Enescu, D. (2019). Thermoelectric energy harvesting: basic principles and applications. *Green Energy Advances*, *1*, 38.
7. Jaziri, N., Boughamoura, A., Müller, J., Mezghani, B., Tounsi, F., & Ismail, M. (2020). A comprehensive review of thermoelectric generators: technologies and common applications. *Energy Reports*, *6*, 264–287.
8. Sandoz-Rosado, E. J., Weinstein, S. J., & Stevens, R. J. (2013). On the Thomson effect in thermoelectric power devices. *International Journal of Thermal Sciences*, *66*, 1–7.

9. Huang, M. J., Yen, R. H., & Wang, A. B. (2005). The influence of the Thomson effect on the performance of a thermoelectric cooler. *International Journal of Heat and Mass Transfer, 48*(2), 413–418.
10. Chen, G., Dresselhaus, M. S., Dresselhaus, G., Fleurial, J. P., & Caillat, T. (2003). Recent developments in thermoelectric materials. *International Materials Reviews, 48*(1), 45–66.
11. Wang, Z. L., Funada, T., Onda, T., & Chen, Z. C. (2023). Knowledge extraction and performance improvement of Bi_2Te_3-based thermoelectric materials by machine learning. *Materials Today Physics, 31*, 100971.
12. Al-Fartoos, M. M. R., Roy, A., Mallick, T. K., & Tahir, A. A. (2023). Advancing thermoelectric materials: a comprehensive review exploring the significance of one-dimensional nano structuring. *Nanomaterials, 13*(13), 2011.
13. Tritt, T. M., & Subramanian, M. A. (2006). Thermoelectric materials, phenomena, and applications: a bird's eye view. *MRS Bulletin, 31*(3), 188–198.
14. Erden, F., Akgul, B., Danaci, I., & Oner, M. R. (2023). Thermoelectric and thermomechanical properties of invar 36: comparison with common thermoelectric materials. *Journal of Alloys and Compounds, 932*, 167690.
15. Kim, H. S., Liu, W., Chen, G., Chu, C. W., & Ren, Z. (2015). Relationship between thermoelectric figure of merit and energy conversion efficiency. *Proceedings of the National Academy of Sciences, 112*(27), 8205–8210.
16. LaLonde, A. D., Pei, Y., Wang, H., & Snyder, G. J. (2011). Lead telluride alloy thermoelectrics. *Materials Today, 14*(11), 526–532.
17. Yamanaka, S., Kosuga, A., & Kurosaki, K. (2003). Thermoelectric properties of Tl_9BiTe_6. *Journal of Alloys and Compounds, 352*(1–2), 275–278.
18. Martin, J., & Nolas, G. S. (2016). Apparatus for the measurement of electrical resistivity, Seebeck coefficient, and thermal conductivity of thermoelectric materials between 300 K and 12 K. *Review of Scientific Instruments, 87*(1), 015105.
19. He, T., Li, X., Tang, J., Zuo, X., Zheng, Y., Zhang, D., & Tang, G. (2020). Boosting thermoelectric performance of BiCuSeO by improving carrier mobility through light element doping and introducing nanostructures. *Journal of Alloys and Compounds, 831*, 154755.
20. Tian, T., Cheng, L., Xing, J., Zheng, L., Man, Z., Hu, D., & Li, G. (2017). Effects of sintering on the microstructure and electrical properties of ZnO-based thermoelectric materials. *Materials & Design, 132*, 479–485.
21. Pumera, M., & Sofer, Z. (2017). 2D monoelemental arsenene, antimonene, and bismuthene: beyond black phosphorus. *Advanced Materials, 29*(21), 1605299.
22. Kurosaki, K., Matsuda, T., Uno, M., Kobayashi, S. I., & Yamanaka, S. (2001). Thermoelectric properties of $BaUO_3$. *Journal of Alloys and Compounds, 319*(1–2), 271–275.
23. Snyder, G. J., & Toberer, E. S. 2008. Complex thermoelectric materials. *Nature Mater, 7*(2), 105–114. https://doi.org/10.1038/nmat2090
24. Sootsman, J., Chung, D., & Kanatzidis, M. 2009. New and old concepts in thermoelectric materials. *Angewandte Chemie International Edition, 48*(46), 8616–8639. https://doi.org/10.1002/anie.200900598
25. Poudel, B., Hao, Q., Ma, Y., Lan, Y., Minnich, A., Yu, B., ... & Ren, Z. 2008. High-thermoelectric performance of nanostructured bismuth antimony telluride bulk alloys. *Science, 320*(5876), 634–638. https://doi.org/10.1126/science.1156446
26. Hsu, K. F. 2004. Cubic $AgPbmSbTe^{2+m}$: bulk thermoelectric materials with high figure of merit. *Science, 303*(5659), 818–821. https://doi.org/10.1126/science.1092963
27. Heremans, J. P, Jovovic, V., Toberer, E. S, Saramat, A., Kurosaki, K., Charoenphakdee, A., ... & Snyder, G. J. 2008. Enhancement of thermoelectric efficiency in PbTe by distortion of the electronic density of states. *Science, 321*(5888), 554–557. https://doi.org/10.1126/science.1159725

28. Dirmyer, M. R., Martin, J., Nolas, G. S., Sen, A., & Badding, J. V. 2009. Thermal and electrical conductivity of size-tuned bismuth telluride nanoparticles. *Small, 5*(8), 933–937. https://doi.org/10.1002/smll.200801206
29. Anderson, M. E, Bharadwaya, S. S. N, & Schaak, R. E. 2010. Modified polyol synthesis of bulk-scale nanostructured bismuth antimony telluride. *Journal of Materials Chemistry, 20*(38),8362. https://doi.org/10.1039/c0jm01424a
30. Mu, E., Yang, G., Fu, X., Wang, F., & Hu, Z. (2018). Fabrication and characterization of ultrathin thermoelectric device for energy conversion. *Journal of Power Sources, 394*, 17–25.
31. Šakalys, R., Mohammadlou, B. S., & Raghavendra, R. (2022). Fabrication of multi-material electronic components applying non-contact printing technologies: a review. *Results in Engineering, 15*, 100578.
32. Liu, Z., Zhang, S., Wu, Z., Mu, E., Wei, H., Liu, Y., ... & Hu, Z. (2023). High-performance integrated chip-level thermoelectric device for power generation and microflow detection. *Nano Energy, 114*, 108611.
33. Abdelkareem, M. A., Mahmoud, M. S., Elsaid, K., Sayed, E. T., Wilberforce, T., Al-Murisi, M., ... & Olabi, A. G. (2022). Prospects of thermoelectric generators with nanofluid. *Thermal Science and Engineering Progress, 29*, 101207.
34. Akinaga, H. (2020). Recent advances and future prospects in energy harvesting technologies. *Japanese Journal of Applied Physics, 59*(11), 110201.
35. Tohid, M. Z. W. B. M., Baharin, T. M. I. B. T., Yatim, H. B. M., Kang, C. C., Suhaimi, F. A. B. M., & Azmi, M. B. (2018). Energy harvester from waste heat using thermoelectricity. *Chemical Engineering, 63*, 781–786.
36. Martín-González, M., Caballero-Calero, O., & Díaz-Chao, P. (2013). Nanoengineering thermoelectrics for 21st century: energy harvesting and other trends in the field. *Renewable and Sustainable Energy Reviews, 24*, 288–305.
37. EnergyNow Media. (2019). Exciting New Environmental Benefits of Thermoelectric Energy Harvesting – Marlow Industries. https://energynow.ca/2019/06/exciting-new-environmental-benefits-of-thermoelectric-energy-harvesting-marlow-industries/#main (Access on 12th August, 2023).
38. Stambouli, A. B. (2011). Fuel cells: the expectations for an environmental-friendly and sustainable source of energy. *Renewable and Sustainable Energy Reviews, 15*(9), 4507–4520.
39. Shi, X., Chen, L., & Uher, C. (2016). Recent advances in high-performance bulk thermoelectric materials. *International Materials Reviews, 61*(6), 379–415.
40. Baxter, J., Bian, Z., Chen, G., Danielson, D., Dresselhaus, M. S., Fedorov, A. G., & Wu, Y. (2009). Nanoscale design to enable the revolution in renewable energy. *Energy & Environmental Science, 2*(6), 559–588.
41. Patyk, A. (2010). Thermoelectrics: impacts on the environment and sustainability. *Journal of Electronic Materials, 39*, 2023–2028.
42. Jang, E., Ambade, R. B., Banerjee, P., Topoleski, L. T., & Madan, D. (2024). Stencil-printed scalable radial thermoelectric device using sustainable manufacturing methods. *Sustainability, 16*(9), 3560.
43. Baroutaji, A., Arjunan, A., Robinson, J., Ramadan, M., Abdelkareem, M. A., Vance, A., ... & Olabi, A. G. (2024). Developments and prospects of additive manufacturing for thermoelectric materials and technologies. *Sustainable Materials and Technologies, 41*, e01008.
44. Jia, S. Z. (2023). Waste energy harvesting in sustainable manufacturing. In Ganesh Naryanan, R. & Gunasekera, J. (eds.). *Sustainable Manufacturing Processes* (pp. 231–256). Academic Press.

2 New Generation Thermoelectric Renewable Energy and Materials Research

2.1 INTRODUCTION

Recrystallization-induced in situ nanostructure formation is a new power generation and solid-state cooling approach for bulk thermoelectric materials [1]. Energy exploration is crucial due to rising demand. Clean, unlimited solar energy is another viable unconventional energy source on Earth and in space. Some solar thermal collectors use thermoelectric generators. System-level theory helps in the operational applications of solar thermoelectric generators (STEG) [2].

To improve PV/TE hybrid device performance, PV modules, TEGs, and cooling systems are added or combined. Prism or spectrum-splitter PV/TE hybrids have perpendicular PV modules and TEGs. Longer wavelengths are reflected to the TEG, while shorter wavelengths are absorbed by the PV module through the splitter (Figure 2.1a). TEGs cool to preserve temperature differences, and PV and TEG process solar energy separately.

PV modules are placed higher, and TEGs lower, because they absorb visible, UV, and IR wavelengths, respectively (Figure 2.1b). Heat sinks cool the cold side of the TEG on the rear. PV modules collect UV and visible solar energy and deliver the rest to the TEG. IR warms the top of the TEG compared to the cool side.

2.2 THERMOELECTRIC MATERIALS AND RENEWABLE ENERGY IN LIMITATION OF CLIMATE CHANGE AND GLOBAL WARMING

A few years ago, experts exaggerated the harm of climate change. Increasing CO2 emissions can be mitigated by using renewable energy. Solar, wind, geothermal, hydroelectric, and piezoelectric energy are renewable. Climate change must be averted by employing all green energy sources [4].

Also concerning is fossil fuel-caused global climate change. Waste heat-scavenging thermoelectric generators power our electronics. Thermal energy from home heating, car exhaust, and industrial processes can power thermoelectrics.

DOI: 10.1201/9781003639343-2

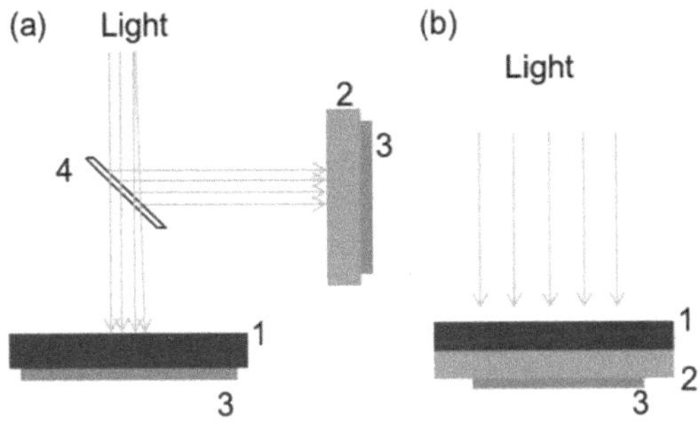

FIGURE 2.1 PV/TE device: (a) reflective component and (b) without reflection—(1) PV module, (2) TEG, (3) heat sink, and (4) reflective component [3].

No-moving-parts thermoelectric generators are silent, reliable, and scalable for distributed power generation. Electrothermic generators replace exhaust alternators for fuel economy [5].

Researchers study alternative energy and energy conversion efficiency, with a focus on thermoelectric (TE) materials. Fast heat-to-electricity conversion makes these materials perfect for sustainable energy. TE materials collect waste heat from renewable energy sources to reduce pollution and improve energy conversion efficiency. An average oil-powered generator is 30% efficient, with wasted heat representing 70% of chemical energy. A TE generator enhances heat-to-electricity efficiency by 2.5%.

TE materials efficiently transform waste heat into power, improving clean energy and reducing greenhouse gas emissions. Reliable devices are used in waste heat recovery, atomic batteries, wearable TE materials, solar-thermoelectric generators, medical sensors, and energy-efficient building glazing. Ineffective TE devices and unsustainable materials prohibit their widespread usage [6].

High-TE materials and structures are being synthesized and manufactured by the Materials Genome Initiative to lower lattice thermal conductivity and improve the TE materials' figure of merit. Low conversion efficiency may benefit parasitic applications. Waste heat or solar energy is cheap or free; hence, TE system efficiency is unimportant.

Automobile manufacturers are building TE generators for waste heat recovery and remote battery replacement. With TE technology, waste heat to power is being considered. TE power units can supply all the electricity in a domestic central heating system for space and water heating.

While electricity is unpredictable, some developing nations use wood or fuel stoves. The technology could be used in commercial autonomous sensors with rechargeable primary batteries and the most expensive components. TE power generators can power low-energy consumers using body heat, and autonomous devices can use both natural and artificial temperatures.

Rare Te, Bi, Sb, and Se are needed for TE. The demand for TE materials may increase as these rare chemicals become scarce. Global warming and environmental impact will decrease with TE technology [7].

Climate change, fossil fuel depletion, and increased energy needs have led to the development of sustainable energy usage, conversion, and recovery technologies. This technology has distinct advantages, including [8]:

1. Since there's no mechanical activity, TE conversion is quieter and more reliable;
2. Eco-friendly because it produces no heat, gaseous, or chemical waste; and
3. In remote areas without energy conversion technologies, it can be used.

2.3 NOVEL APPROACHES IN THERMOELECTRIC RENEWABLE ENERGY CONVERSION

In 1914, an experimental and patented thermoelectric generator recovered waste heat from a reciprocating engine (Figure 2.2). Exhaust fumes heat the pipe, and cooling water heats the heat sink. Each circulation path receives cooling water from pumps. A pipe on the exhaust pipe circulates the cooling water. For this purpose, the cooling water, as indicated in Figure 2.2, displays no mechanical movement.

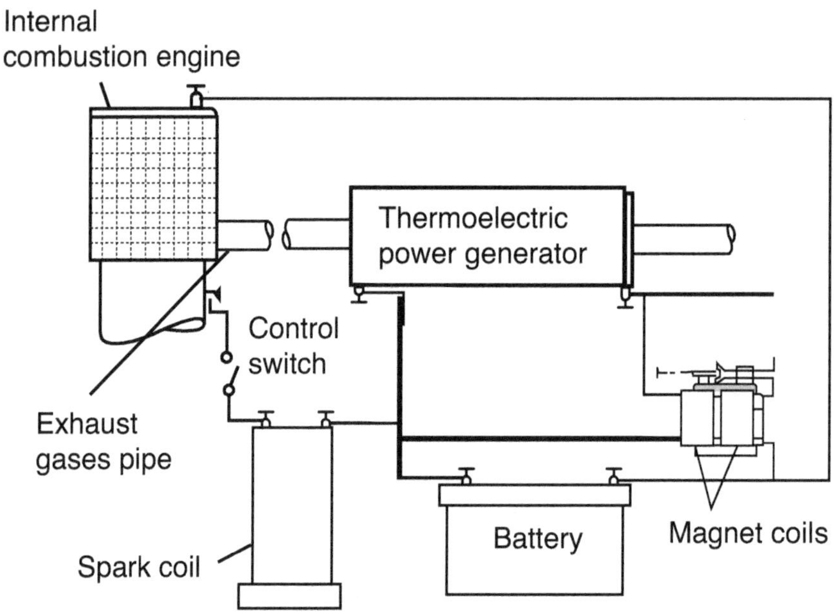

FIGURE 2.2 Schematic of early thermoelectric power generator-based waste heat-to-electricity conversion for internal combustion engines [9].

Cold thermoelectric generator sides are cooled by pump water. A circulation loop with enough cooling water could increase the temperature difference. A cooling water pump requires extra energy for circulation. When analyzing this "enhancement" from water pumps, consider the energy necessary to circulate liquid and extract heat at the cooling end. Moreover, 50% of most cars' petrol energy is lost as exhaust and radiator heat, with less than 25% efficiency.

Recent research has studied hybrid PV-TE systems. Figure 2.3 shows essential schemes. Using light and heat efficiently is crucial. PVs use visible and UV light, while TEs use IR (Figure 2.3a). Figure 2.3b displays the hybrid PV-TE system. The open circuit voltage and power output are 78 mV and 0.19 μW, respectively. Evaluations included 10% TEG-powered hybrid systems. In the analysis, the TEG junction temperature difference linearly affected current, voltage, and efficiency (Figure 2.3c). Hybrid solar systems used PV, TEG, concentrator, and heat engine. During the summer and winter months, a photovoltaic-thermal (PVT) heat pipe-driven TE heating and cooling system was conducted for theoretical and experimental analysis. PVT has 23.5% thermal and 16.7% electrical efficiency. A sophisticated heat transfer model investigated PVT-TE power generation. To evaluate load matching and performance, a thermodynamic hybrid PV-TE system was created. Integrating PV with air source heat pumps was effective.

A hybrid Dye-Sensitized Solar Cell-Zinc Oxide (DSSC-ZnO) TE nanogenerator generates electricity from light and mechanical energy (Figure 2.3d). After the Indium Tin Oxide (ITO) glass and oxide layers undergo a chemical reaction, sunlight is able to travel through the dye and Au layers, resulting in the generation

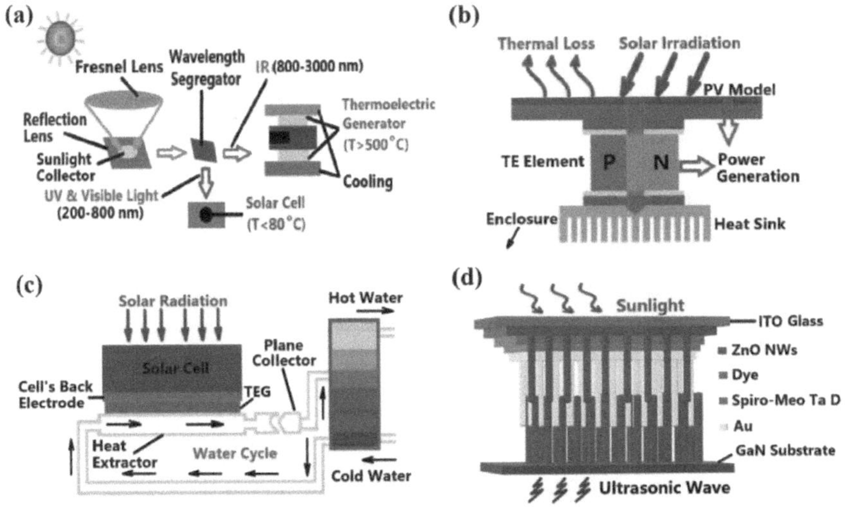

FIGURE 2.3 (a) Solar energy captured in concentrated form, (b) graphical representation of a hybrid PV–TE system, (c) structure of a hybrid photovoltaic-thermal (PV/T) system with thermoelectric generators (TEGs) and nonconcentrated radiation, and (d) an illustration of the structure of a compact hybrid cell [10].

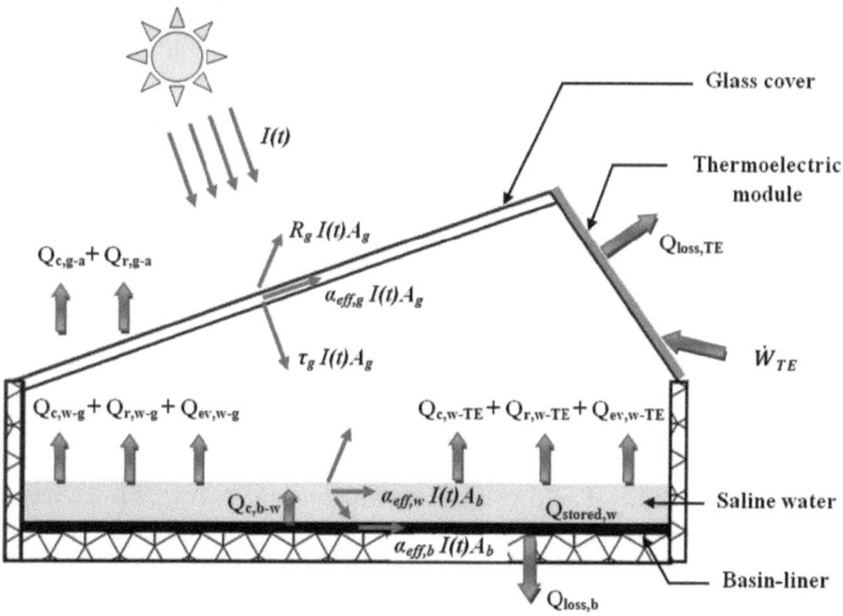

FIGURE 2.4 A single-basin solar still featuring a thermoelectric module's main energy transfer systems [11].

of photovoltaic electricity. This process is analogous to the P-N junction. As a result of the temperature differential, the hot and cold ends of the TE module are made of Spiro-MeO, and they have a common end with a GaN substrate layer. High energy conversion is challenging due to the low efficiency of hybrid systems. Light absorption, light splitting, material selection, construction, cooling, and optimization are some of the possibilities that researchers have proposed for improving the system [10].

The solar system must be smaller for portability. Production, solar radiation, and the sizes of the condensing and evaporating zones decrease. A 12 V DC-powered thermoelectric module (TEC1–12708) cools the inside to increase the evaporating-condensing gradient (Figure 2.4). Models predicted a productivity of 0.25 L/m²/day and 7% energy efficiency. Thermoelectric module solar stills produced no better results.

A total of three thermoelectric modules have been added, the thermoelectric surface has been angled, the clear water collecting plate has been removed, and the angle of the glass cover has been adjusted. To identify the quantity, position, and causes of irreversibilities, the thermal system and its components must undergo an exergetic performance study. The exergy destruction rate in each component of a passive single-slope solar still has been explored as an alternative to the new solar still (SS) that incorporates thermoelectric modules. Most solar still system research disregards saline water exergy for simplicity, but it is considerable and should be included in exergy balancing [11].

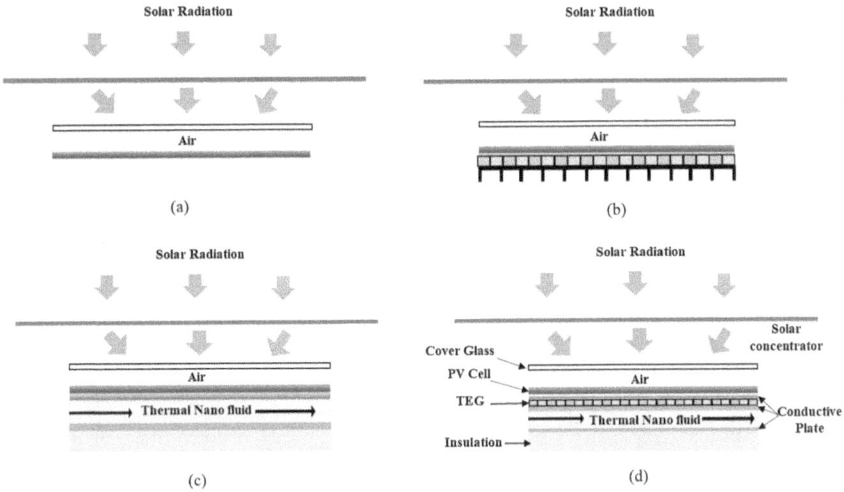

FIGURE 2.5 Detailed description of the body: (a) using a standard CPV, (b) hybrid CPV/TEG-HS control system, (c) the hybrid system of NCPV/T, and (d) proposed NCPV/T-TEG hybrid system [12].

Figure 2.5a–c compares the proposed new design's performance with other configurations. Figure 2.5d depicts the physical nanofluid-based concentrated photovoltaic/thermal-thermoelectric generator (NCPV/T-TEG) system. Most of this research found that a TEG's cold side heat sink enhanced PV cell-module heat transfer. Heat sinks help with TEG and PV cell cooling. Conversion energy loss reduces system efficiency. Liquids cool faster than heat sinks in a gearbox. The Solar/T-TEG hybrid uses waste heat with nanofluid. The heat sink is replaced by a nanofluid coolant. Solar and TEG performance are improved, and waste heat is used to create electricity [12].

This hybrid device's PV cell may use photoelectricity (Figure 2.6a). Heat is wasted through the solar cell's front surface radiation and convection. Finally, TE modules receive the PV cell's heat and radiation. The TE module utilizes all of the solar energy absorbed by the PV cell. As a result of the Seebeck effect, TE modules are able to generate power while simultaneously cooling. Figure 2.6b depicts the structure of a solar cell with a single period [13].

Figure 2.7 shows the concentrated photovoltaic-thermoelectric generator (CPV-TEG) power and Thermostatic and Expansion Valve (TEV) systems in the suggested TEV. Photovoltaic and thermoelectric effects power the CPV-TEG power system solely by solar energy. The TEV system is used for constructed ventilation, including winter and summer load supply, and would be electric. Without it, heating and cooling would use more energy. Changing the input current direction switches TEV heating/cooling. Concentrating sunlight increases energy density. Light-energy conversion to thermal energy may increase the PV panel temperature (Figure 2.8). Several buttoned TEGs can fully cycle thermal

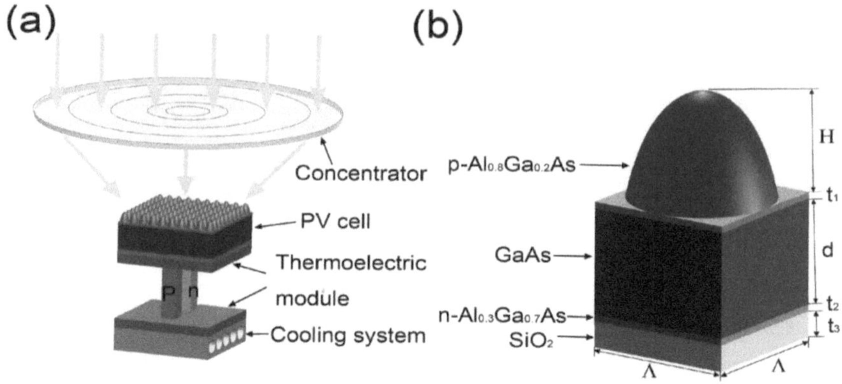

FIGURE 2.6 (a) The PV–TE hybrid system with an optical concentrator is depicted in this schematic, and (b) within a single frame, the structure of the photovoltaic cell [13].

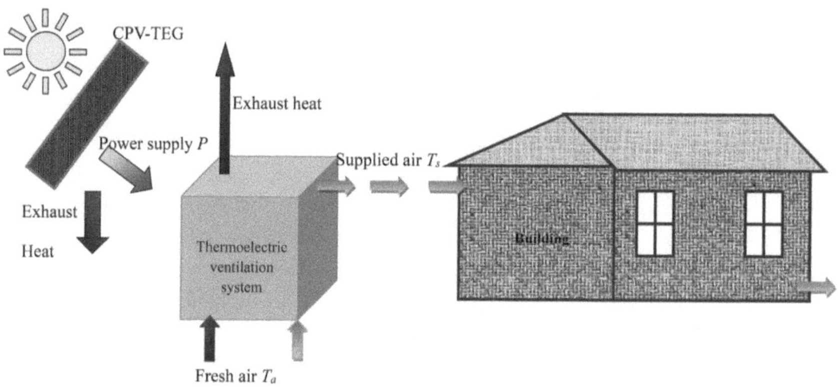

FIGURE 2.7 Energy-efficient building concept with concentrated photovoltaic-thermoelectric generator-driven thermoelectric ventilation system [14].

energy to generate electricity. Power and reliability are better with CPV-TEG than with a single PV system [14].

A single-slope, single-basin SS and external condenser are used. All basins, SSs, and external condensers are made of 1 mm galvanized material. Basins are 50 cm long and wide, so their effective area is 0.25 m^2. The schematic of the external condenser and nanofluid system is shown in Figure 2.9. The glass cover frame is secured to the SS with 9 bolts, 0.01m EPS, and thermal silicone adhesive. An exterior mirror on the SS reflects sunlight. Thermoelectric modules are installed on the basin bottom using silicon adhesive.

A front-still one-way valve collects water. The double-slope symmetrical condenser has roof basins for product water. The condenser's bottom has two valves to collect water from both sides. The sides of the double-slope condenser are

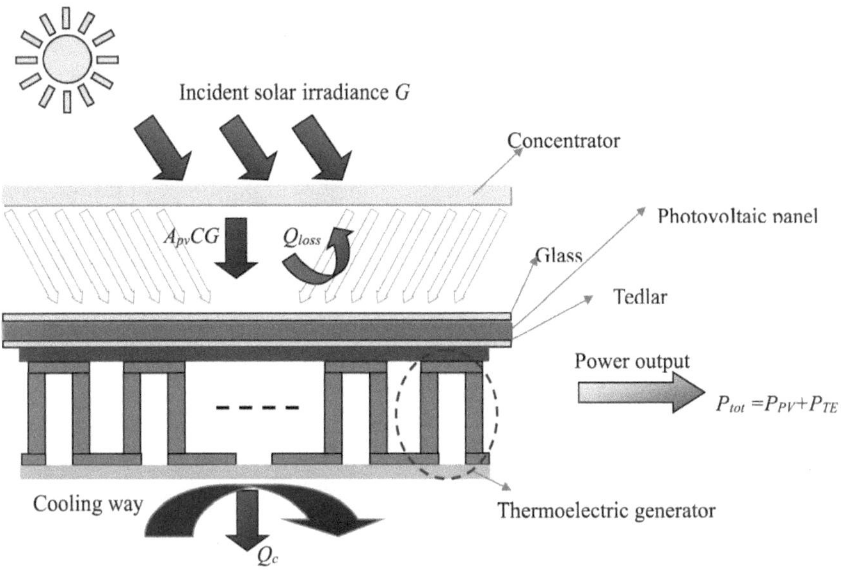

FIGURE 2.8 Diagram of CPV-TEG electricity supply system [14].

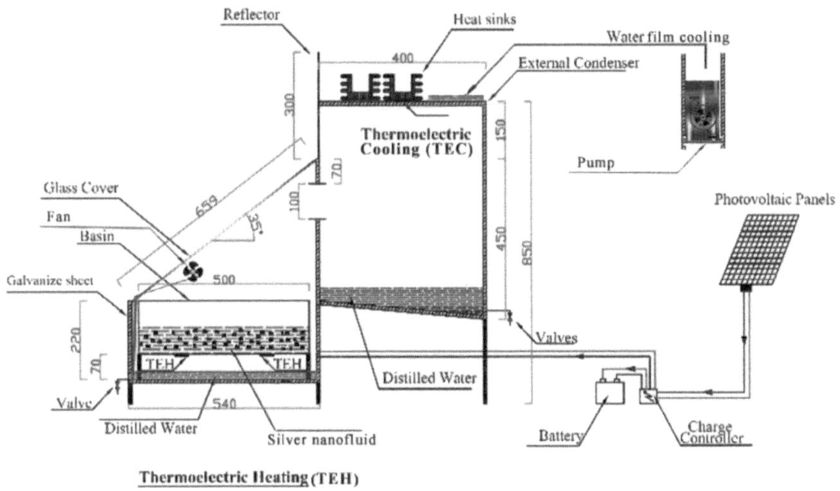

FIGURE 2.9 A diagrammatic representation of the system, which includes an external condenser and nanofluid [15].

cooled differentially. The right condenser roof has two cold-side 12,704 thermoelectric modules. Each module has hot-side aluminum heat sinks. The heat sinks reduce the module's hot-side temperatures, extending its life. Pharmaceutical firm reverse osmosis (RO) desalination water cools the left roof [15].

2.4 THERMOELECTRIC POWER GENERATION SYSTEMS

Figure 2.10 shows the test rig schematic diagram. The TEG block has 16 four-layer thermoelectric modules with coolers and heaters. A Bi_2Te_3 TEM has 210 40 mm × 3.8 mm thermoelectric modules. Materializing solid surfaces with thermal interfaces improved heat transfer. The hot side had two 100 W controlled cartridge heaters. Flowmeters measured the cooling water flow, and K-type thermocouples measured the heat source and heat sink temperatures. The controller and thermocouple accuracy affected the heat flux temperature set-point inaccuracy. Inaccurate thermocouple temperature indicators and set-point temperature overshoot caused temperature controller errors [16].

The TEG system's adjustable external load was rheostat resistance from 0 to 40 Ω. Instrumental errors cause experimental test results to differ from predictions. Voltage and current issues limit power. The accuracy of the controller and thermocouple affected the heat flux temperature set-point. Set-point temperature overshoot and thermocouple temperature indicator inaccuracy caused temperature controller errors (Figure 2.11).

In Figure 2.12, the open-circuit electric potential of the two TEMs is depicted under a variety of parameters for the engine. Since the two curves are parabolic, this gap increases with decreasing inward heat input. This analysis differs from others with constant material properties because semiconductors are temperature-dependent.

Heat contact is reduced by TM pressure, enhancing system performance. In Figure 2.13, fluid flow rates affect power output. Increased cold fluid flow lowers

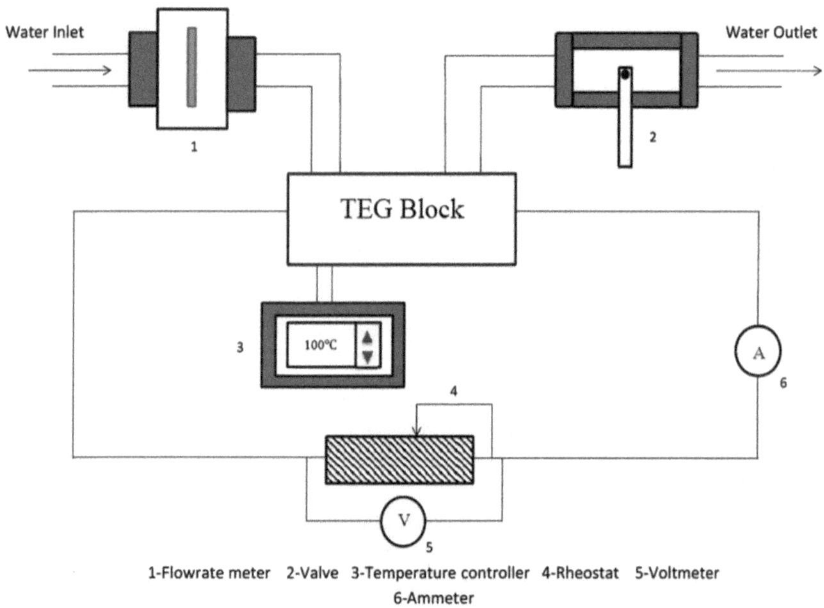

FIGURE 2.10 An outline of the testing apparatus [16].

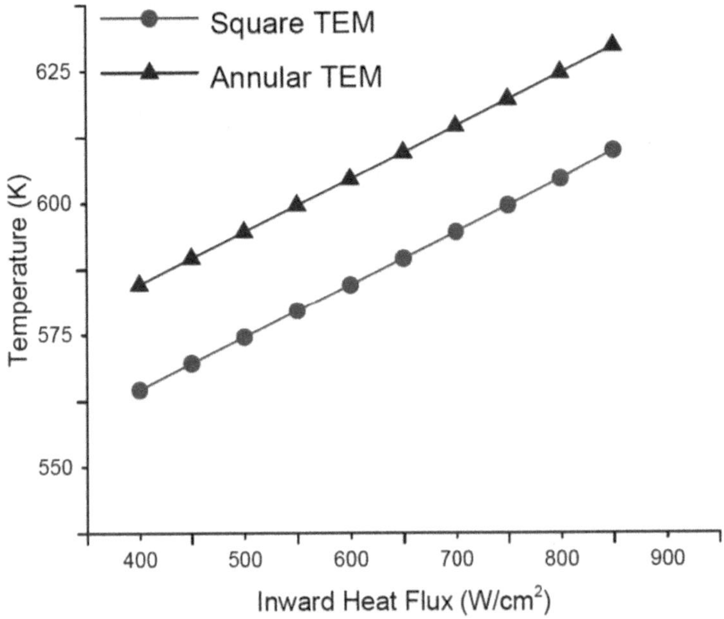

FIGURE 2.11 Both TEM hot-side average temperatures are variable depending on the amount of inward heat flux [16].

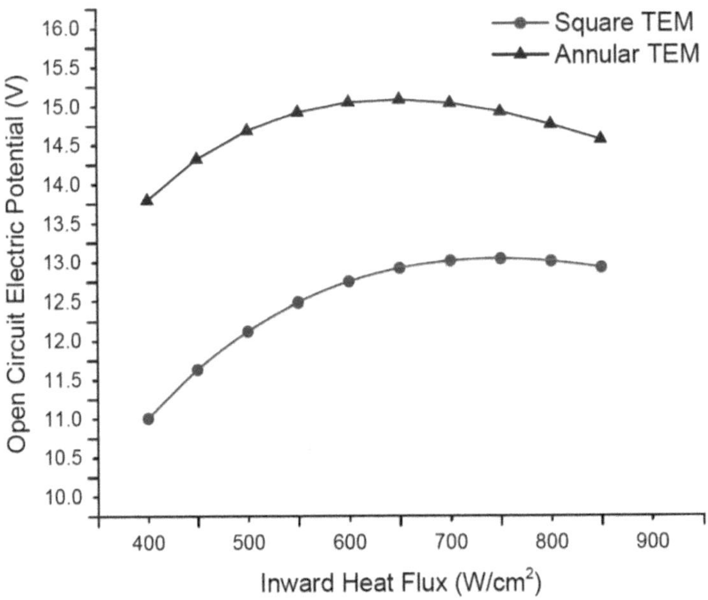

FIGURE 2.12 Electric potential of a dual TEM open circuit applied to a variety of inward heat fluxes [16].

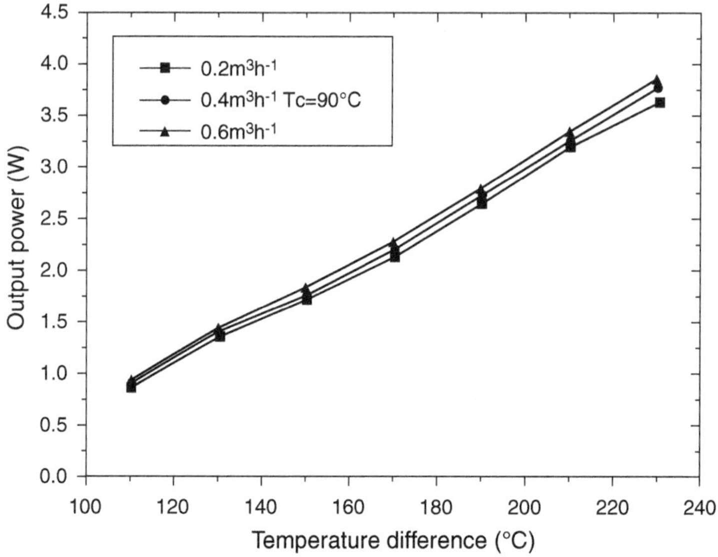

FIGURE 2.13 The influence of cold-side fluid flow on the variation of power output [17].

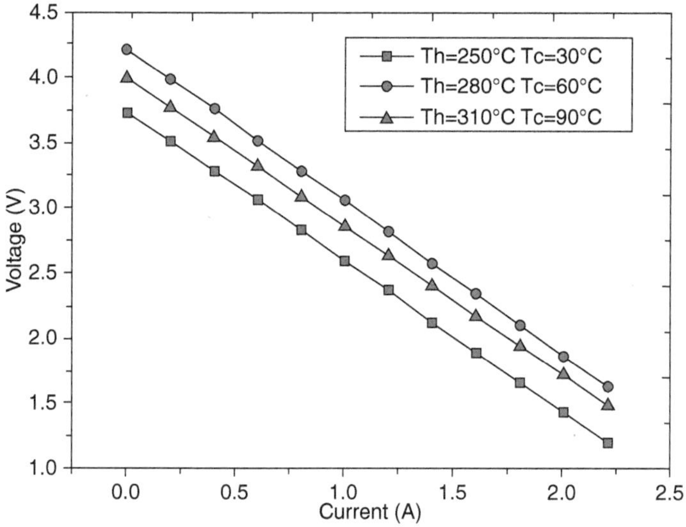

FIGURE 2.14 Output voltage–current curves for three situations [17].

the TM cold-side temperature. The engine cooling system should be upgraded with a water-cooling unit in order to save space and maintain the cold-side temperature at 90°C (Figure 2.14) [17].

Develop and construct thermoelectric systems for experimental purposes. There are eight Bi_2Te_3 thermoelectric modules that measure 200mm by 160mm by

12.64 mm. An electronic load (FAST AUTO ELECTRONIC FA2300) with 1% current and 0.1% voltage precision confirmed the thermoelectric system's performance. Experimenting at different temperatures increases the open voltage as the temperature difference grows (Figure 2.15). The operating temperature was 505 K, with a maximum open voltage of 59 V. Maximum power production and thermoelectric conversion efficiency increase with the temperature difference (Figure 2.16).

The predicted conversion efficiency trend matches the experimental power output, with 10.9% and 2.8% conversion efficiencies achieved at 505 K and 75

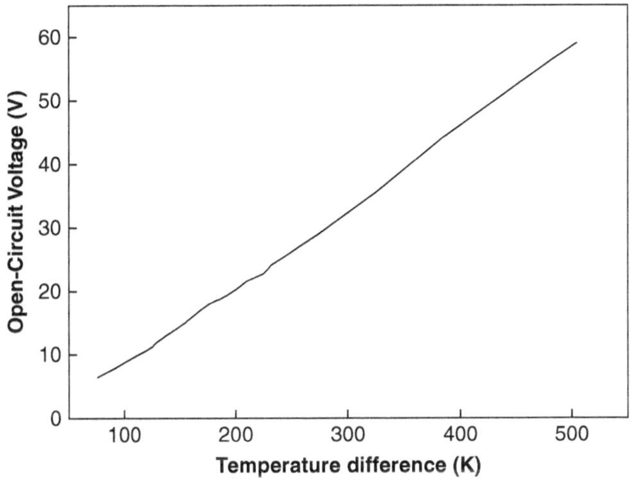

FIGURE 2.15 The open circuit voltage output is dependent on the operating temperature [18].

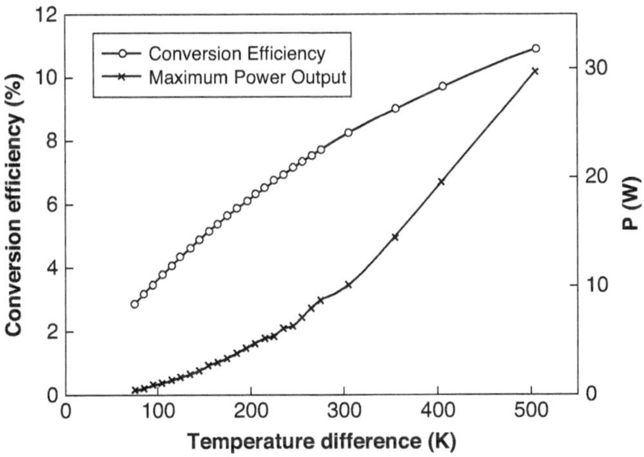

FIGURE 2.16 Depending on the temperature, the efficiency of thermoelectric conversion and the maximum power output [18].

K, respectively (Figure 2.17). Power output is more affected by TH and TC than by voltage. These findings suggest that a higher TH will provide more power and may not change the voltage at the same temperature [18].

A test system of the cryogenic fluid-based cold-energy thermoelectric power-generating experimental setup is shown in Figure 2.18. Nitrogen in cryogenic tanks is in a gas–liquid state under pressure. The nitrogen temperature rises with the tank pressure. The storage tank's liquid nitrogen transfers heat with the open precooling tank cryogenic liquid (–196°C).

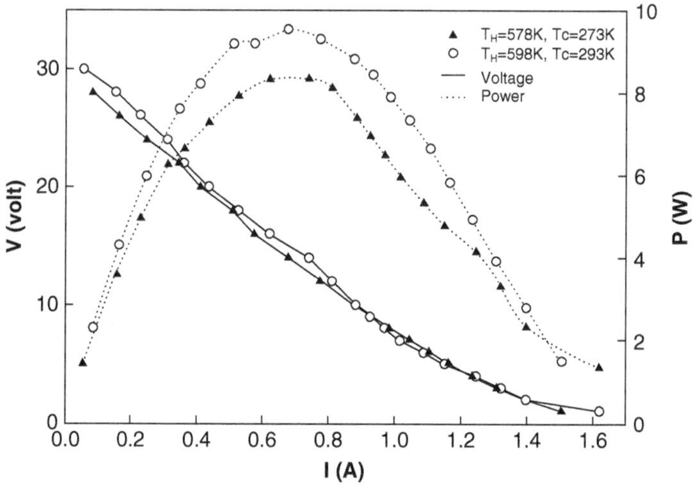

FIGURE 2.17 The V–I and P–I curves correspond to varying T_H and T_C values when the temperature difference is the same [18].

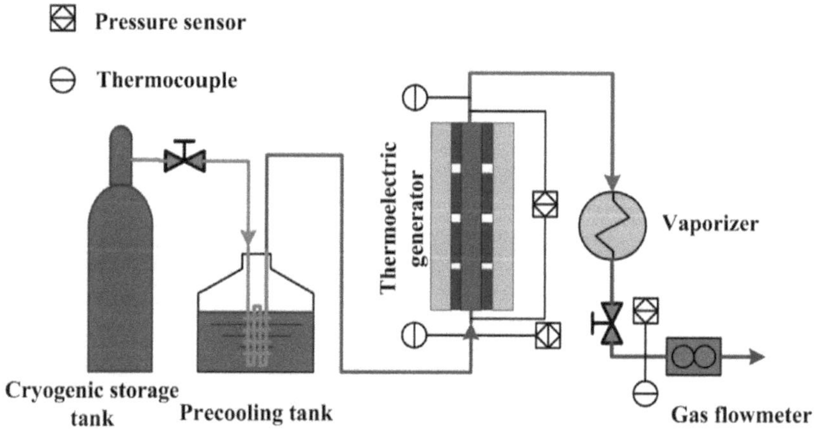

FIGURE 2.18 An illustration of the cold energy thermoelectric generator system in schematic form [19].

Liquid nitrogen enters the TEG subcooled. Cold liquid nitrogen gasifies, heating the TEG. The TEG generates power from the liquid nitrogen-heat source temperature differential. Gas or gas–liquid nitrogen is passed by the TEG. Cryogenic nitrogen cannot enter the gas flowmeter due to its low temperature. The flow rate is measured after vaporizer heating. A vacuum insulation pipe connects the pre-cooler and low-temperature storage tank to prevent cold energy loss in the pipeline.

The LN2 heat exchanger, thermoelectric modules, and finned exchanger are all components of the cold-energy TEG setup shown in Figure 2.19. In the liquid nitrogen and finned heat exchangers, the contact thermal resistance was

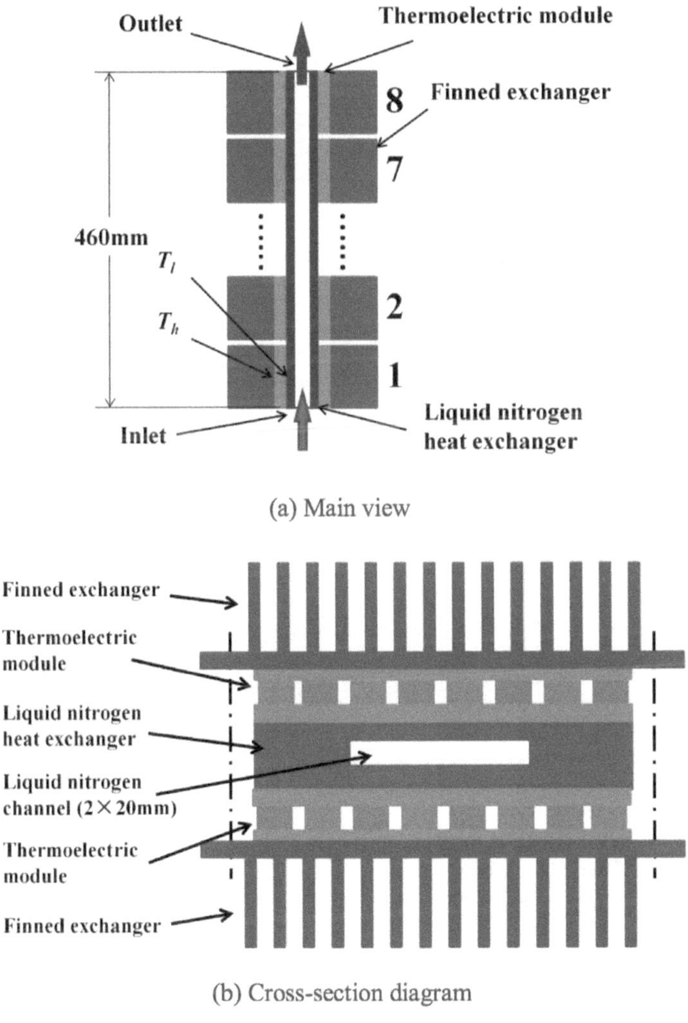

FIGURE 2.19 A diagrammatic representation of the thermoelectric generator for cold energy [19].

decreased by coating the cold and hot surfaces of the module with graphite that was 0.5 mm thick.

Each heat sink screw was tightened to 0.3 kg·m. The air-heated gasifier warms the DEG. Both the air-side heat transfer and the hot-side temperature of the thermoelectric modules are improved using aluminum fin exchangers. Both the height and thickness are 45 mm, while the spacing is 3 mm.

Easy installation and better thermoelectric module temperature dispersion are achieved with the flat liquid nitrogen heat exchanger. The liquid nitrogen flow channel has a 2 mm × 20 mm cross-section and is 460 mm long with a 10 mm by 56 mm cross-section [19].

2.5 EMERGING TRENDS AND FUTURE PERSPECTIVES

Build "hydro-electro-thermal" maps by correlating neutron imaging, local current density, and temperature. Humidity, oxygen, and heat flow data from other researchers were unique. These measurement sets are acceptable for neutron imaging but have not been integrated with imaging, which warrants future investigation. Figure 2.20 shows Psi's Peltier thermo-electric cooler cell for heat flow measurement and control. The anode-cathode heat flow distribution, lab-scale cold and subzero starting, and evaporative cooling are also depicted. In this advanced setup (Figure 2.20), neutron imaging through Bismuth Telluride thermoelectric coolers examined cell water distribution [20].

2.5.1 Development of New Thermoelectric Materials

A smaller hot-cold temperature difference improves thermoelectric air cooling. Modules with a greater Seebeck coefficient among thermoelectric materials with comparable merit are better for moderate temperature cycling [21].

Despite substantial research, thermoelectric materials are difficult to apply. Increasing the output of high-performance thermoelectric materials is essential to increase conversion. Improving the chemical, thermal, and mechanical stability of thermoelectric materials is also necessary. High Seebeck coefficient materials enhance output power and thermoelectric performance. The Multipurpose Ion Generator (MGI) and high-throughput computing and synthesis have assisted thermoelectric material research. Theories and technologies from scientists should advance thermoelectric research [22].

2.6 PROMISING DIRECTION OF FUTURE RESEARCH AND KEY FINDINGS

Many organic and hybrid organic–inorganic thermoelectrics function poorly. To develop next-generation thermoelectric materials, doping, thermal, and electrical transport mechanisms must be understood in complex systems with multiple components and both ordered and disordered phases. Optimization of varied processing may require understanding structure-property interactions

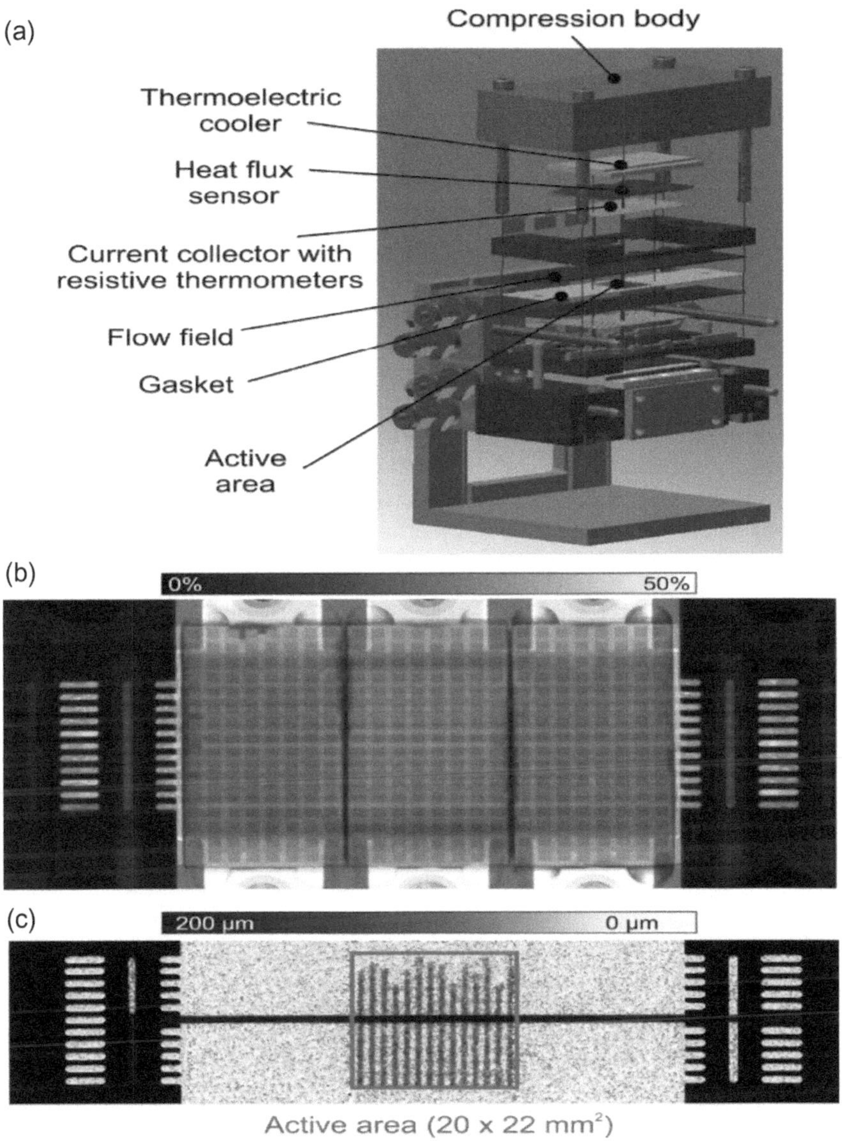

FIGURE 2.20 Experiment cell that is compatible with neutron imaging and features improved temperature management: (a) cell construction; (b) through the structure of the cell, transmission occurs; and (c) iterative measurement of water distribution [20].

between the semiconducting polymer and the predicted composite components. Semiconducting polymers with carbon nanotube (CNT) interactions can be used in composite thermoelectrics when organic semiconductors are tuned for thermoelectrics.

Thin polymer films and other organic thermoelectric materials are measured differently than inorganic batch products. Conventional material device groups report Seebeck coefficients and in-plane thermal conductivities based on environmental circumstances, sample size, shape, and substrate. The diverse shapes and topologies of organic thermoelectric devices make their performance unstandardized. Power factor and mechanical properties are being evaluated for new thermoelectric composites. Innovative, mechanically flexible thermoelectric device designs and applications enhance efficiency. Using growing material classes to find innovative thermoelectric device applications would maximize their distinctive properties. Metallic and covalent organic frameworks, gels, foams, and conjugated microporous polymers are examples of crystalline frameworks. Thermoelectric materials and devices have many unexplored avenues [23].

In the pursuit of high-performance thermoelectric materials, trial-and-error selection of materials and synthesis conditions often fails. Computational research can explain essential atomic-scale properties and the thermoelectric effect in new materials without requiring extensive experimental characterization. Computer simulations can forecast reliable new thermoelectric material designs [24].

2.6.1 Emerging Trends and Future Perspectives

TE materials based on p-type PbTe have higher ZT values than n-type due to their band structure and delicate alterations. N-type PbTe can increase its Seebeck coefficient and power factor by creating a resonant impurity level or dispersing low-energy carriers through potential barriers at interfaces. Reducing thermal conductivity is an effective way to raise ZT.

For this, bulk nanostructured TE materials with phonon scattering across all wavelengths seem promising. Middle-wavelength phonons, which transfer up to 50% of thermal energy, scatter most effectively with short- and long-wavelength phonons. Nanopowder spark plasma sintering produces bulk nanostructured TE materials with ZT values at 98%–99% of theoretical density.

High ZT values result from an integrated approach and the simultaneous action of numerous factors to enhance the power factor and decrease thermal conductivity, which together increase the thermoelectric figure of merit of the TE material [25].

Microelectronics have become smaller and more adaptable in recent decades, resulting in smart devices like handheld phones, cardiac pacemakers, and exoskeletons that have revolutionized our lifestyles by changing their energy sources. Power supply is a major difficulty in electronic device design and implementation as energy consumption per unit of space decreases.

Innovative wearable TE energy conversion devices generate electricity from body heat to power portable devices. Although still in development, flexible film- and fiber-based TE energy conversion devices possess outstanding chemical, physical, and mechanical qualities and could be employed in portable low-power electronic power sources.

Many advanced film- or fiber-based TE energy conversion devices have been tested for low-power wearable electronics. The primary achievements are as follows:

1. Optimization of variable preparation techniques for components, structure, and post-processing conditions and times has generated various high-performance flexible film- and fiber-based TE materials.
2. Flexible film- and fiber-based energy conversion devices have incorporated waste heat in prototypes. Human body heat can power TE energy conversion devices, promising sustainable electricity for wearable electronics.
3. Film- or fiber-based TE converters provide wearable energy collection potential. By using high-performance flexible film or fiber materials for p- and n-type materials, many wearable devices can generate renewable power that supports low-power electronics.

Film- and fiber-based materials are ideal for flexible or wearable TE energy conversion devices, but conversion efficiency, durability, and safe electrical power supply systems still require further development. The following points are considered:

1. Enhance the evaluation of film- or fiber-based materials and devices. The four-point probe and laser flash/three-point methods, along with the development of an energy detection system, assess thermal and electrical conductivity under ideal conditions. Recent investigations show the TE performance of film- or fiber-based materials; however, measurement variation and energy loss are unavoidable due to adverse conditions and unequal cross-sections.
2. Several film- or fiber-based wearable devices have output performance that is too low for commercial use. Wearable energy converters should be able to power hearing aids, pacemakers, biosensors, etc., with more than 5 W.
3. Film- and fiber-based legs for TE energy conversion devices should be more stretchable since people move and stretch or compress.
4. Designing a commercial prototype by employing 3D printing or multiple energy sources.

Wearable energy harvesting for commercial deployment may involve 3D printing fine micro TE conversion devices to embed in clothing, weaving them into fabric to provide electronic energy for low-power or embedded electronics using novel wireless charging technology, or manufacturing them. Wearable TE energy conversion devices will be key interdisciplinary issues as flexible and integrated electronics become more widely available [26].

Mass production is coming for mature and reliable TE power-generating module fabrication technologies. The findings reveal that TE devices are generating more power. Sizing and process optimization are needed to increase conversion efficiency and minimize module production costs.

The device must be more reliable at high temperatures and oxygen levels for practical use. Study service behavior under difficult operating settings (vibration, oxidation, thermal cycling, etc.). The lack of accurate measurement methods and mature, reliable instrumentation for TE element and module evaluation is another obstacle. A life cycle and operation reliability prediction technique should consider the electrical, mechanical, and thermal performance of materials and devices.

Finally, many areas must collaborate to comprehend the rational design, cost-effective production, and evaluation of optimum TE materials and modules to advance TE technology's implementation in numerous power-generating applications [27].

Most wearable thermoelectric generator research has concentrated on choosing suitable materials and perfecting the modules' thermoelectric legs' structural and geometrical form. Thermal design, heat sink geometry optimization, thermoelectric generator placement on the body, free and forced heat convection, and the effects of temperature and humidity on thermoelectric generator module performance are also studied.

Significant research has been done on wearable thermoelectric generators to power stand-alone wireless sensors, wireless sensor networks, and wearable devices; however, there are still gaps that can be addressed to speed up product development.

Several studies have assessed the performance of thermoelectric generators that have been used over time at various temperatures. However, due to these variations, it is currently challenging to compare their performance accurately and identify the optimal material and construction. The investigation of enhancing the flexibility of wearable thermoelectric generators to minimize contact resistance with the skin has received considerable attention. However, the techniques for connecting the modules to the skin surface and the optimal contact pressure have not been thoroughly examined.

During regular use, modules may not adhere to the skin, thereby impacting performance, although most research assumes optimal contact between the thermoelectric module and the heat source. Sweating impacts the heat resistance of the skin's thermoelectric surface. The impact of perspiration on the thermoelectric performance of wearable generators has not been investigated; hence, further research is required. The majority of studies on wearable thermoelectric generators operate under the assumption of a consistent skin temperature.

Physical activity, metabolic rate, skin tissue structure in various body areas, thermal resistance, clothing factor, garment disparity, ambient temperature, relative humidity, and air velocity all have an impact on skin temperature. To tackle this problem, certain studies have utilized the Pennes bioheat equation to calculate skin temperature.

This technique disregards the body's thermoregulatory responses to external temperature cues, such as skin vasodilation, vasoconstriction, and perspiration. Utilizing thermoregulatory models for the calculation of skin temperature has the potential to enhance the accuracy of forecasts. No changes were made to the user's text. Power converters that are unable to electrically match thermoelectric generators (TEGs) negatively affect the performance of waste-heat thermoelectric generators (WTEGs) in simulations [28].

REFERENCES

1. Shen, J. J., Zhu, T. J., Zhao, X. B., Zhang, S. N., Yang, S. H., & Yin, Z. Z. (2010). Recrystallization induced in situ nanostructures in bulk bismuth antimony tellurides: a simple top down route and improved thermoelectric properties. *Energy & Environmental Science, 3*(10), 1519–1523.
2. Jarman, J. T., Khalil, E. E., & Khalaf, E. (2013). Energy analyses of thermoelectric renewable energy sources. *Open Journal of Energy Efficiency, 2*(4), 143–153.
3. Huen, P., & Daoud, W. A. (2017). Advances in hybrid solar photovoltaic and thermoelectric generators. *Renewable and Sustainable Energy Reviews, 72*, 1295–1302.
4. Snyder, G. J., & Toberer, E. S. (2008). Complex thermoelectric materials. *Nature Materials*, 7(2), 105–114.
5. Zhu, T., Liu, Y., Fu, C., Heremans, J. P., Snyder, J. G., & Zhao, X. (2017). Compromise and synergy in high-efficiency thermoelectric materials. *Advanced Materials, 29*(14), 1605884.
6. Al-Fartoos, M. M. R., Roy, A., Mallick, T. K., & Tahir, A. A. (2023). Advancing thermoelectric materials: a comprehensive review exploring the significance of one-dimensional nano structuring. *Nanomaterials, 13*(13), 2011.
7. Tesfaye, F. (2018). An overview of advanced chalcogenide thermo-electric materials and their applications. *Journal of Electronic Research and Application*, 2(2), 28–41.
8. Han, C., Li, Z., & Dou, S. (2014). Recent progress in thermoelectric materials. *Chinese Science Bulletin, 59*, 2073–2091.
9. Zheng, X. F., Liu, C. X., Yan, Y. Y., & Wang, Q. (2014). A review of thermoelectrics research – recent developments and potentials for sustainable and renewable energy applications. *Renewable and Sustainable Energy Reviews, 32*, 486–503.
10. Tang, J., Ni, H., Peng, R. L., Wang, N., & Zuo, L. (2023). A review on energy conversion using hybrid photovoltaic and thermoelectric systems. *Journal of Power Sources, 562*, 232785.
11. Dehghan, A. A., Afshari, A., & Rahbar, N. (2015). Thermal modeling and exergetic analysis of a thermoelectric assisted solar still. *Solar Energy, 115*, 277–288.
12. Lekbir, A., Hassani, S., Ab Ghani, M. R., Gan, C. K., Mekhilef, S., & Saidur, R. (2018). Improved energy conversion performance of a novel design of concentrated photovoltaic system combined with thermoelectric generator with advance cooling system. *Energy Conversion and Management, 177*, 19–29.
13. Da, Y., Xuan, Y., & Li, Q. (2016). From light trapping to solar energy utilization: a novel photovoltaic–thermoelectric hybrid system to fully utilize solar spectrum. *Energy, 95*, 200–210.
14. Cai, Y., Wang, W. W., Liu, C. W., Ding, W. T., Liu, D., & Zhao, F. Y. (2020). Performance evaluation of a thermoelectric ventilation system driven by the concentrated photovoltaic thermoelectric generators for green building operations. *Renewable Energy, 147*, 1565–1583.
15. Parsa, S. M., Rahbar, A., Koleini, M. H., Aberoumand, S., Afrand, M., & Amidpour, M. (2020). A renewable energy-driven thermoelectric-utilized solar still with external condenser loaded by silver/nanofluid for simultaneously water disinfection and desalination. *Desalination, 480*, 114354.

16. Huang, K., Li, B., Yan, Y., Li, Y., Twaha, S., & Zhu, J. (2017). A comprehensive study on a novel concentric cylindrical thermoelectric power generation system. *Applied Thermal Engineering*, *117*, 501–510.
17. Liu, X., Li, C., Deng, Y. D., & Su, C. Q. (2015). An energy-harvesting system using thermoelectric power generation for automotive application. *International Journal of Electrical Power & Energy Systems*, *67*, 510–516.
18. Ma, H. K., Lin, C. P., Wu, H. P., Peng, C. H., & Hsu, C. C. (2015). Waste heat recovery using a thermoelectric power generation system in a biomass gasifier. *Applied Thermal Engineering*, *88*, 274–279.
19. Ge, M., Li, Z., Wang, Y., Zhao, Y., Zhu, Y., Wang, S., & Liu, L. (2021). Experimental study on thermoelectric power generation based on cryogenic liquid cold energy. *Energy*, *220*, 119746.
20. Boillat, P., Lehmann, E. H., Trtik, P., & Cochet, M. (2017). Neutron imaging of fuel cells–recent trends and future prospects. *Current Opinion in Electrochemistry*, *5*(1), 3–10.
21. Duan, M., Sun, H., Lin, B., & Wu, Y. (2021). Evaluation on the applicability of thermoelectric air cooling systems for buildings with thermoelectric material optimization. *Energy*, *221*, 119723.
22. Cai, B., Hu, H., Zhuang, H. L., & Li, J. F. (2019). Promising materials for thermoelectric applications. *Journal of Alloys and Compounds*, *806*, 471–486.
23. Finn, P. A., Asker, C., Wan, K., Bilotti, E., Fenwick, O., & Nielsen, C. B. (2021). Thermoelectric materials: current status and future challenges. *Frontiers in Electronic Materials*, *1*, 677845.
24. Gutiérrez Moreno, J. J., Cao, J., Fronzi, M., & Assadi, M. H. N. (2020). A review of recent progress in thermoelectric materials through computational methods. *Materials for Renewable and Sustainable Energy*, *9*, 1–22.
25. Shtern, Y., Sherchenkov, A., Shtern, M., Rogachev, M., & Pepelyaev, D. (2023). Challenges and perspective recent trends of enhancing the efficiency of thermoelectric materials on the basis of PbTe. *Materials Today Communications*, *37*, 107083.
26. Li, C., Jiang, F., Liu, C., Liu, P., & Xu, J. (2019). Present and future thermoelectric materials toward wearable energy harvesting. *Applied Materials Today*, *15*, 543–557.
27. Zhang, Q. H., Huang, X. Y., Bai, S. Q., Shi, X., Uher, C., & Chen, L. D. (2016) Thermoelectric devices for power generation: recent progress and future challenges. *Advanced Engineering Materials*, *18*(2), 194–213.
28. Tabaie, Z., & Omidvar, A. (2023). Human body heat-driven thermoelectric generators as a sustainable power supply for wearable electronic devices: recent advances, challenges, and future perspectives. *Heliyon*, *9*(4), e14707.

3 Design and Manufacturing of Thermoelectric Devices

3.1 INTRODUCTION

In Thermoelectric Generator (TEG), energy conversion efficiency depends on the thermoelectric device's thermal integration. Thermal integration lowers the thermoelectric module-thermal reservoir temperature differences, improving system efficiency. Research work on TEGs includes energy conversion efficiency optimization of thermoelectric device architecture and materials. TEG efficiency depends on energy conversion system integration. Effective cold and hot plate heat transfer, low thermal resistance, and low heat leakage improves efficiency. To maximize efficiency, the design should consider the integrated system's thermal arrangement [1].

Better TE materials make device development and construction important. Device evaluation requires a complete process, unlike TE materials, where a single quantity figure of merit (zT) may indicate performance best. Due to lower thermal conductivity, which increases the temperature gradient and thermal expansion-induced internal stress, TE material fails mechanically below a certain leg length. In 2D quantum wells, 1D quantum wires, and zero-dimensional quantum dots, electron charge carrier quantum confinement improves the Seebeck coefficient and power factor. Many surfaces scatter phonons more than electrons, decreasing heat conductivity more than electrical conductivity. This fundamental semiconductor had low lattice thermal conductivity and high ZT because copper ions behaved like liquids around a Se crystalline sublattice in $Cu_{2x}Se$.

A high-efficiency "phonon-liquid electron-crystal" thermoelectric material was created using a crystalline sublattice for electronic conduction and liquid-like ions. ZT values around 1.0 are best for commercial thermoelectrics. Other high-performance thermoelectric materials exhibit ZT values of 1.2–2.2 at 600–800 K. Similar to home refrigeration, ZT 4 devices may attain 30% Carnot efficiency [2,3].

Most applications use tellurium and germanium, making many thermoelectric materials costly. Effective materials were generated quickly using nanostructuring and new crystal architectures. Current prototypes show thermoelectric devices' terrestrial, scalable uses. Cost and energy conversion performance determine scalable thermoelectric device development and implementation. The most promising thermoelectric materials' production and system costs must be assessed

DOI: 10.1201/9781003639343-3

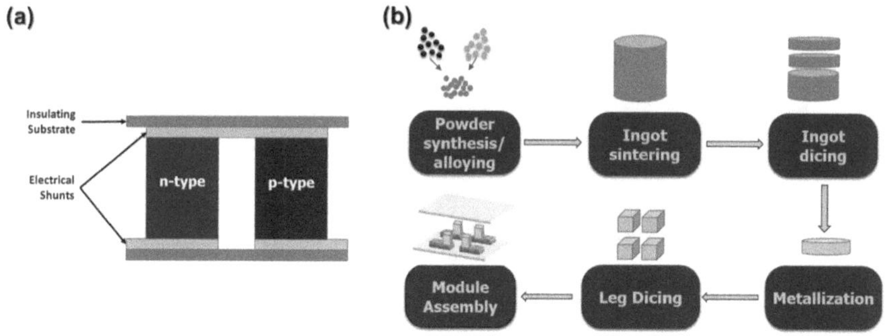

FIGURE 3.1 (a) This thermoelectric device has numerous couplings between two electrically insulating surfaces and (b) conventional thermoelectric device assembly [5].

immediately for commercial viability. Material extraction costs are included in battery and PV cost-performance studies. Cost is typically overlooked in thermoelectric device-linked thermal and electrical module performance evaluations [4].

Thermoelectric devices generate electricity from heat. The Seebeck effect occurs when energetic electrons from a semiconductor thermoelectric material's hot side spread to the cold side until an electric field stops them. P- and N-leg thermocouples are electrically in series and thermally in parallel (Figure 3.1a). Materials development has enhanced ZT as waste heat recovery thermoelectric power generation expands. Very high material ZT does not inhibit device integration. Poor material processing and integration still make thermoelectric generators less efficient and more expensive (Figure 3.1b).

3.2 THERMOELECTRIC DEVICES DESIGN CONSIDERATIONS

The temperature difference (Δtm) between TE couplings in a thermoelectric power generator with a Peltier module and two heat exchangers was estimated by equating conduction heat and heat transfer from hot to cold fluid. Heat exchanger measurements indicate that the heat transfer coefficient and flow rate impact the temperature difference (Δtm).

3.2.1 Heat Transport in Fluids

The following energy conservation equations approximate one-dimensional fluid heat transport. The convection of exhaust gas heat to the TE modules' hot side is described by [6],

$$m_e\left(C_{pe,j}T_{e,j}-C_{pe,j-1}T_{e,j-1}\right)-U_{e,j}A_e\left(T_{e,j}-T_{H,j}\right)=0 \tag{3.1}$$

Heat converts from the cold side of TE modules to the coolant using this equation:

$$m_f\left(C_{pf,j}T_{f,j}-C_{pf,j+1}T_{f,j+1}\right)-U_{f,j}A_f\left(T_{L,j}-T_{f,j}\right)=0 \tag{3.2}$$

where

Subscripts e and f = Exhaust gas and coolant,
$C_{pe,j}$ and $C_{pf,j}$ = Specific heat capacity of the fluids,
$U_{e,j}$ and $U_{f,j}$ = Average heat transfer coefficient of the modules,
$T_{e,j}$ and $T_{f,j}$ = Temperatures of the fluids,
$T_{e,j} = \left(T_{e,j} + T_{e,j+1}\right)/2$ and $T_{f,j} = \left(T_{f,j} + T_{f,j+1}\right)/2$,
$T_{H,j}$ and $T_{L,j}$ = Temperatures at the hot and cold sides of TE module,
m_ e and m_ f = Mass flow rates of the fluids, and
A_e and A_f = Heat transfer area between the fluid and the TE module.

This applies to minor ΔT situations. The case assumes constant heat exchanger thermal resistances. Conduction and forced convection are approximable, but not radiation and natural convection. To maximize power in low ΔT applications, the thermoelectric module's thermal resistance should equal the sum of all external resistances. The ideal thermoelectric design parameters for heat exchangers with a small ΔT are expanded to include linear and sinusoidal thermal resistance [7].

Low and medium-low temperature TGMs are commercially available, while medium-high to high temperatures are rare. The nuclear-powered American Voyager spacecraft's thermoelectric generators used extensively doped n- and p-type silicon–germanium alloy thermocouples. For decades, radioisotope thermoelectric generators have worked. Application requires system-level TGM integration, making TGMs intermediary products. The development of new waste heat recovery thermoelectricity markets requires high-efficiency thermoelectric materials, generator modules, heat conversion systems, and module-application integration. When designing the module, consider thermal conductivity and application [8].

3.2.2 Requirements of Thermoelectric Property Materials

Important factors when reporting material thermoelectric properties are as follows:

- Electrical and thermal properties should be measured in the same direction to avoid anisotropic bias.
- Measuring heating and cooling cycles is recommended for phase transitions in the temperature range of interest.
- A thermoelectric device requires equal thermal and mechanical properties for n- and p-type materials. The device may not work at either component's correct zT temperature.
- To compare reports across device designs and settings, use $W/cm^2{\cdot}K^2$ and $W/g{\cdot}K^2$ for power and $V/cm^2{\cdot}K$ and $V/g{\cdot}K$ for voltage [9].

Wearable inorganic TE systems started with metal, a novel material. More options include nanotubes, graphene, carbon fibers, fullerenes, and carbon nanomaterial/conducting polymer composites. These materials are mechanically flexible but thermoelectrically inefficient. Flexible bulk thermoelectrics are another option.

Inorganic materials become flexible by depositing nanostructures on flexible surfaces or adding PEDOT:PSS. Despite weak flexibility, inorganic films on flexible substrates have an excellent power factor. Because of their higher TE, organic/inorganic nanocomposite films should contain as much inorganic material as possible. Due to their chemical differences, composites with perfect interfaces between the phases are challenging to make. All-inorganic flexible TE materials have been created and used more quickly due to the plastic deformability of bulk Ag_2S and InSe, which may be explored. High-TE, flexible candidate materials have improved [10].

In addition, thermoelectric devices can precisely and responsively manage dynamic temperature. Micro-thermoelectric coolers (μTECs) provide quick and accurate localized temperature control. Micro-TECs regulate telecommunication laser diode temperatures and manage on-chip hot spots, achieving precise cooling in high heat flux locations (600–1300 W/cm^2). Combining TECs with additional cooling stages can improve performance in packages. Various technologies, including electrodeposition, sintering, thin film, and microsystems technology, have been employed to generate μTECs. Cost-effective, scalable, ambient-operable electrochemical deposition is compatible with CMOS back-end technology for μTECs on a chip [11].

Optimized thermoelectric harvesters: TEGs that directly utilize local temperature differences can treat the energy source as a constant-temperature heat stream. Only the source structure's finite thermal conductance affects the TEG's input temperature. Maximal energy output requires maximal heat flow and TEG efficiency. Direct thermoelectric harvesters function best with a TEG temperature differential of $\Delta T/2$, similar to power transfer load matching [12].

Full factorial parametric and Latin hypercube model parameters are determined by general research. First, a manifold model with pre- and post-fin inlet and output volumes is examined. Its easy manufacturing and steady water volumetric flow distribution are key features. The complex fin origin and termination geometry rapidly adds mesh pieces as the volume expands. As the mesh size increases, computing time also increases without changing temperature profiles. Fins without heatsink height and a top gap do not lower cold side module temperatures and increase computational cost due to more complex flow. According to simulations, fins should reach heatsink height [13].

Low-dimensional material production is costly and yields low. Researchers have proposed nanobulk structuring, band convergence, and resonant levels to increase zT in chalcogenides, skutterudites (SKUs), half-Heusler (HH) compounds, certain silicides, silicon–germanium alloys, and segmented TE legs with the correct temperature range. TE materials outpace device research. Device performance requires a holistic approach, unlike TE materials, where zT can show performance. Temperature gradient and thermal expansion stress rise with low thermal conductivity, restricting the mechanical failure of TE material below a certain leg length. Enhance the power factor and thermal conductivity. Nanocrystalline silicon and half-Heusler compounds are preferable [2].

Solid-state thermoelectric generators work well. Advanced thermoelectric devices can replace power-generating heat engines in lightweight heat recovery systems. Maximum thermoelectric energy conversion efficiency depends on heat reservoir temperature and zT. Many materials with high zT values are created to improve thermoelectric efficiency. A good dimensionless figure-of-merit is zT/degree, where T (K) is the average operational temperature. Limit TE devices' power conversion efficiency. Bismuth Telluride (Bi_2Te_3) was the sole commercial thermoelectric module material with 1.35 and 0.9 p- and n-type figure-of-merit values. N-type Bismuth Telluride-based materials have lower TE unit efficacy than p-type due to reduced productivity. Bi_2Te_3 and alloys have a 180–450 K useful temperature range for TE refrigeration and low-power niche applications. Spacecraft power production uses PbTe and SiGe, which can tolerate 500–900 K [and 800–1300 K] [14].

3.3 MODELING AND SIMULATION IN THERMOELECTRIC DEVICE DESIGN

TEG performance and structural optimization employ theory. Model predictions save time and money over experiments, especially when optimizing several parameters. TEG and TEG system performance must be predicted using credible models. A Seebeck voltage is generated when p-type thermoelectric holes and n-type electrons migrate from hot to cold in the TEG [15].

The Peltier effect causes carrier accumulation at the cold side interface to release heat, while dissipation at the hot side absorbs heat. Due to their temperature differential, thermally conductive semiconductors generate Thomson heat, while electric current, metal electrodes, and load resistance generate Joule heat. These important concepts have been used to create TEG models for analysis and optimization.

Basic models include thermal resistance, numerical, and analogy. It cannot be expanded; the one-dimensional thermal resistance model only considers heat transfer from hot to cold.

The following TEG analysis methods are available:

3.3.1 Thermal Resistance Models

The hybrid transient CFD-thermal resistance model uses both. This hybrid transient model requires the following:

- The Thomson effect's small impact is ignored.
- Solving the thermal resistance model assumes homogenous TEM surface temperatures.
- TEMs with equal thermal grease on both sides and proper ATEG system clamping remove structure-to-structure thermal resistance. The early calculation reveals that this ignorance causes a 0.26% inaccuracy, minimally affecting ATEG system performance.

Figure 3.2 presents the schematic of the hybrid transient CFD-thermal resistance model. The 1/4 ATEG system consists of four TEMs connected in series, resulting in identical currents flowing across them. The TEMs responsible for the exhaust flow are TEM1, TEM2, TEM3, and TEM4.

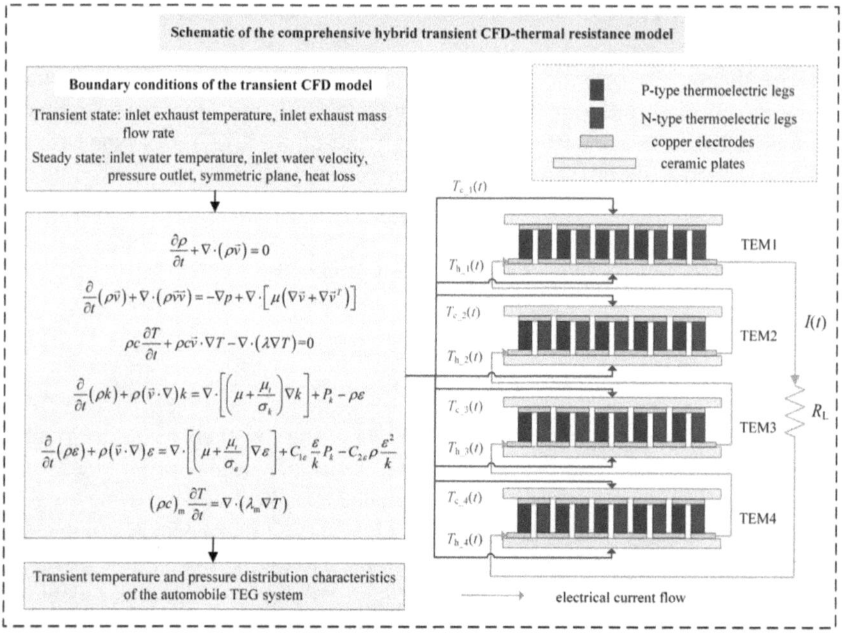

FIGURE 3.2 The entire transient hybrid CFD-thermal resistance mode [15].

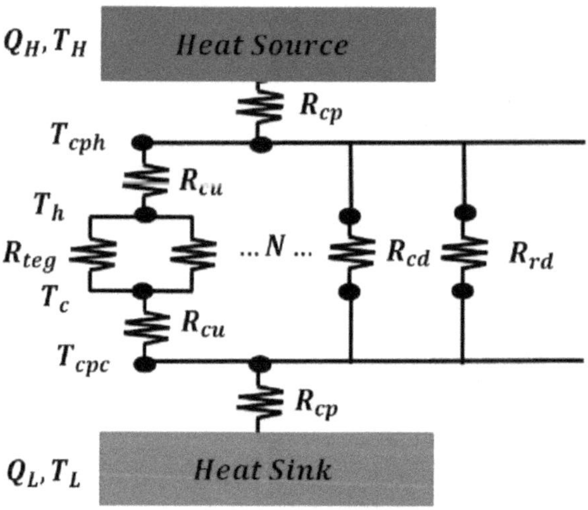

FIGURE 3.3 Schematic of TEG device's thermal resistance network [16].

3.3.1.1 The Simplified Model

A thermal resistance network produces the thermal loss equations for the entire model in Figure 3.3. Radiative and conduction heat loss are calculated by the thermal resistance network. Low Rayleigh numbers reduce convection and damage. For temperature gradients ΔT of 570 K, 470 K, 370 K, and 270 K, the

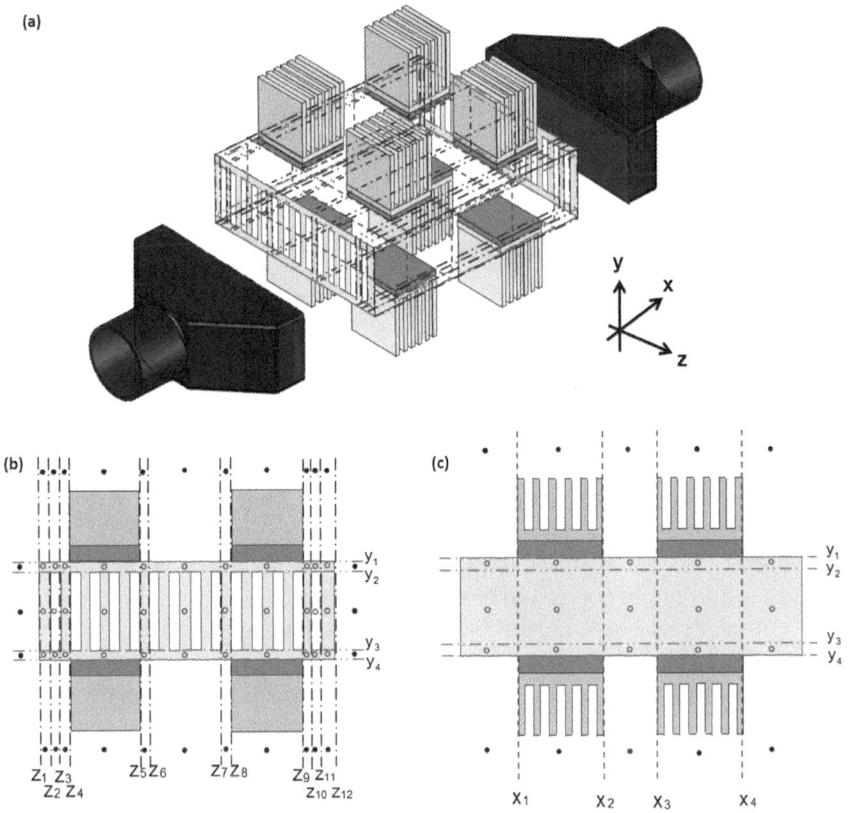

FIGURE 3.4 Thermal resistance model for waste heat recovery. (a) System-wide domain; (b) Front system view; (c) System side view [17].

comprehensive 1-D model has Rayleigh numbers of 1298, 1070.27, 842.56, and 614.84. Air gap heat radiation and conduction are indicated by all four Rayleigh numbers being below 1700 because natural heat convection cannot occur.

In Figure 3.4, the resistance R_{cp} and R_{cu} are given as [16]:

$$R_{cp} = \frac{\delta_{cp}}{k_{cp} A_{cp}} \tag{3.3}$$

$$R_{cu} = \frac{2\delta_{cu}}{k_{cu} A} \tag{3.4}$$

where

δ_{cp} = Thickness,
k_{cp} = Thermal conductivity,
A_{cp} = Surface area,

k_{cu} = Thermal conductivity, and
A = Cross-sectional area.

The packing fraction (θ) is used to describe the air-gap geometry in TEG.

$$\theta = \frac{n\left(A_p + A_n\right)}{A_{cp}} \tag{3.5}$$

For p–n legs having the same dimensions, and the electrical resistance of Copper is

$$r_{cu} = \frac{\rho_{cu}\left(1 + \frac{1}{\theta}\right)}{\delta_{cu}} \tag{3.6}$$

3.3.1.2 The Improved Thermal Resistance Model

The ambient heat sinks, TEG modules, and the waste heat recovery chamber are all part of the computational domain in Figure 3.4a. Along with the system input and output outside, these cells have temperature nodes at their centroid. The same ambient temperature (T_{inf}) boundary conditions are represented by black nodes in Figure 3.4b and c. In contrast to conventional designs, which require convective heat transfer coefficients, this cuboid chamber method recovers heat.

This 3D thermal resistance analysis follows:

The domain comprises thirteen z-axis sections (z_1, z_2, z_3,..., z_{13}) and five y-axis sections (y_1, y_2, y_3, y_4, and y_5). Using TEG placements, Figure 3.4c divides the domain into five x-axis portions at x_1, x_2, x_3, x_4, and x_5.

1. It locates cell centroids with unknown temperature nodes. The TEG matches heat input and output in open-loop circuits to conserve cell energy. Figure 3.5 depicts the thermal resistance relationship between nodes and neighbors.

$$\frac{T_j - T_{fr}}{R_{th,fr}} + \frac{T_j - T_e}{R_{th,e}} + \frac{T_j - T_w}{R_{th,w}} + \frac{T_j - T_n}{R_{th,n}} + \frac{T_j - T_s}{R_{th,s}} = 0 \left(\text{for inlet nodes}\right) \tag{3.7}$$

$$\frac{T_j - T_{br}}{R_{th,br}} + \frac{T_j - T_e}{R_{th,e}} + \frac{T_j - T_w}{R_{th,w}} + \frac{T_j - T_n}{R_{th,n}} + \frac{T_j - T_s}{R_{th,s}} = 0 \left(\text{for outlet nodes}\right) \tag{3.8}$$

3.3.2 Numerical Models

The L-TEG with p/n-type TE legs and electrodes is shown in Figure 3.6. Electricity can now flow when RL closes the L-TEG circuit. Joule heating and Fourier's heat conduction are involved in TE conversion. The process is ongoing and internal.

Using these assumptions simplifies the issue:

1. Adiabatic L-TEG walls. Do not consider radiation or convection heat.
2. Contact resistances are neglected because the electrode-TE leg interface is perfect.
3. P- and n-type legs are similar [18].

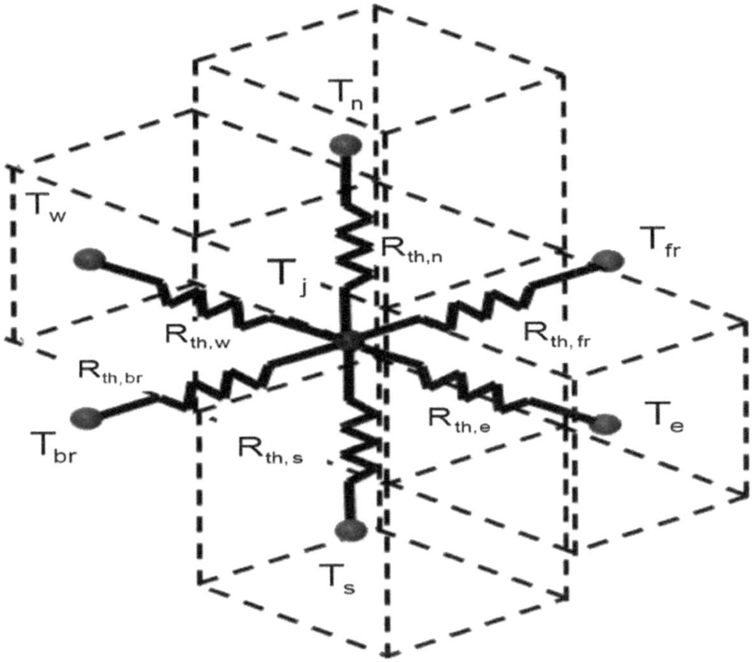

FIGURE 3.5 A relationship between node (Tj) and six others (Tbr, Tfr, Te, Tw, Tn, Ts) using thermal resistance concept [17].

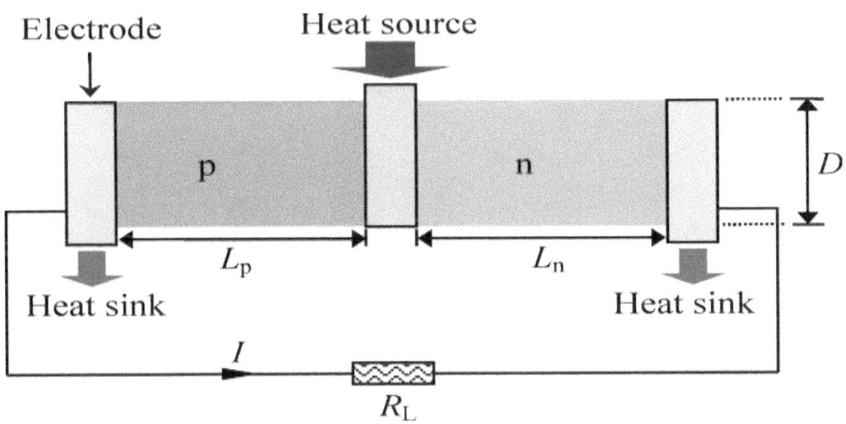

FIGURE 3.6 Schematic diagram of the L-TEG [18].

Exhaust gas transient and steady-state boundary conditions are used to solve the transient CFD model of the automobile TEG system (Figure 3.7). The transient thermal-electric numerical model of all TEMs is solved using the grounded boundary condition to obtain the dynamic electric response of the vehicle TEG system [19].

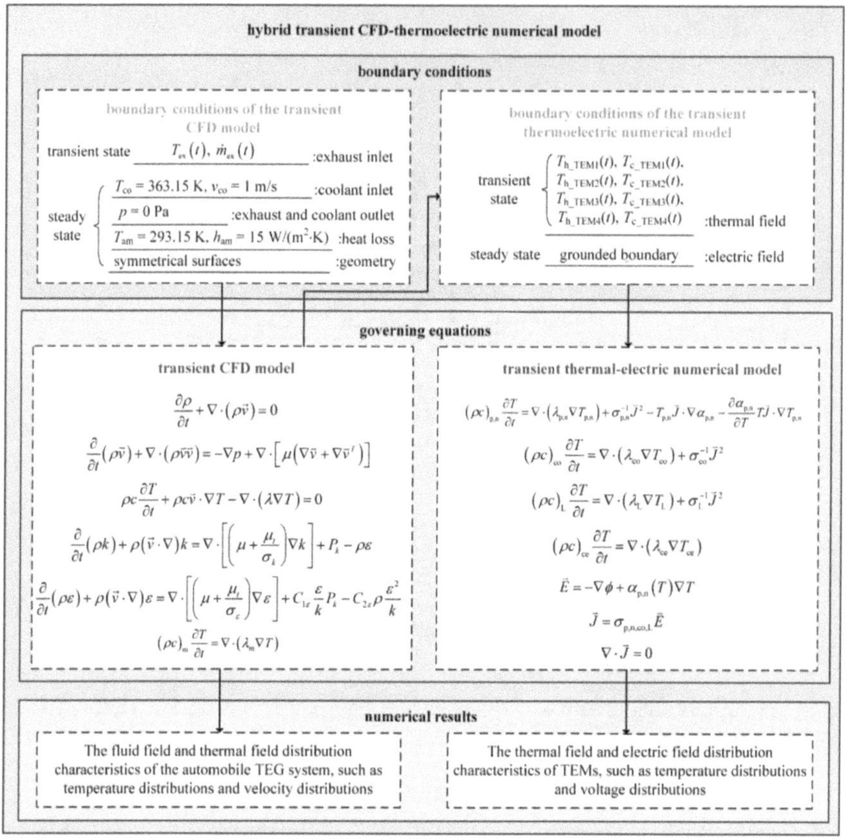

FIGURE 3.7 The hybrid transient CFD-thermoelectric numerical model's governance equations and boundary conditions [19].

3.3.3 Comparison of Different Models

Recently, TEG device theoretical models have made considerable strides. Researchers have developed more extensive theoretical models for TEG units, modules, steady states, and transients. There are numerical, analog, and thermal resistance models. Numerical model computation is slow. Thermal resistance models are simpler and solve faster than numerical models, but their error is larger.

The dynamic response of the TEG device under pulse input can be studied using an analogous model for electronic control circuit simulation. The use of the thermal resistance model's core equations causes errors in the analogy model. The numerical model is the most commonly used TEG device performance prediction model because it properly depicts physical field distribution [20].

The 3D model uses current instead of load resistance. Figure 3.8 displays the TEG unit schematic and transient model boundary conditions. Steady-state

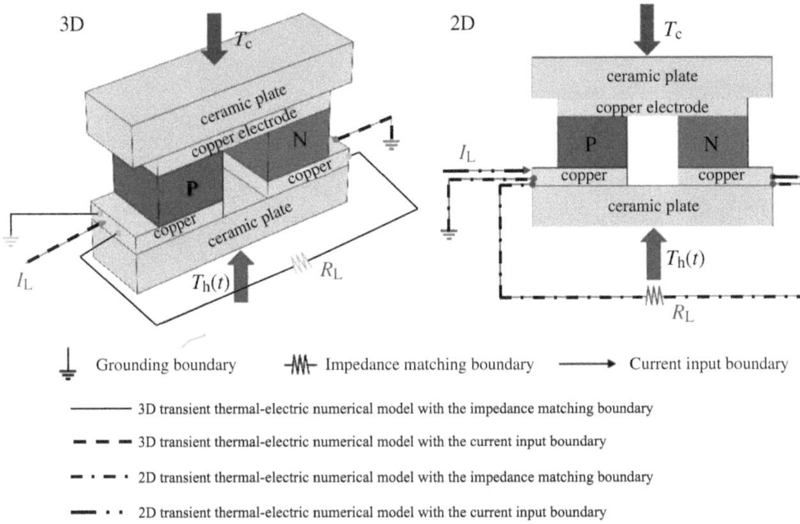

FIGURE 3.8 Different transient model boundary conditions and TEG unit diagram [21].

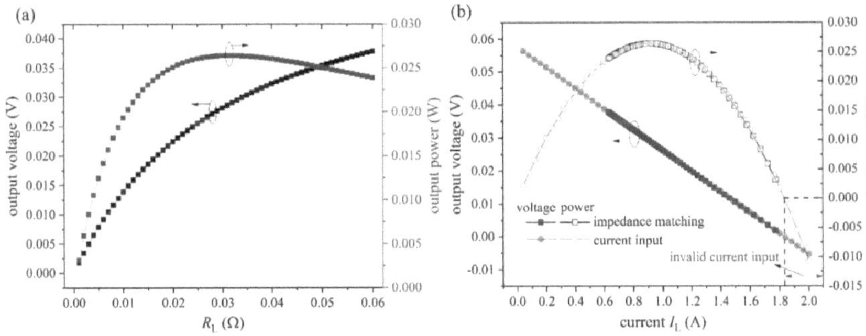

FIGURE 3.9 (a) The 3D model's impedance-matched boundary output prediction. (b) 3D model output performance with current input boundary [21].

numerical simulations preceded the selection of load resistance and current input. The 3D and 2D models have identical optimal working positions. Figure 3.10 displays two 3D models predicting the TEG unit's steady-state output at a 450 K heat source temperature. The TEG unit output power peaks at R_L=0.31 Ω and I_L=0.92 A (Figure 3.9a and b).

The model's steady-state output performance is the same with impedance matching and current input bounds (Figure 3.9b). Too much input current encourages carrier directional movement in thermoelectric materials instead of a temperature differential, resulting in poor output.

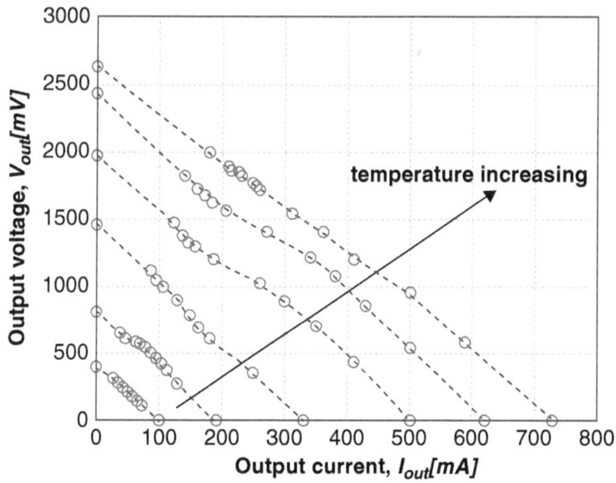

FIGURE 3.10 The relationship between the output voltage, V_{out} [mA], and the output current, I_{out} [mA], is measured at various test temperatures, T_{test} [°C] [24].

3.4 MANUFACTURING TECHNIQUES FOR THERMOELECTRIC DEVICES

Printing polymer TE devices affects development and performance. Doped semiconductors and semimetals dominate TE research. Chalcogenides, silicides, clathrates, skutterudites, half-Heusler alloys, oxides, and tetrahedrites are TEs. Many local heating/cooling devices use bismuth telluride-based TEs. Commercializing TE devices is difficult due to the scarcity, toxicity, high raw material costs, and processing challenges of Bi, Te, Sb, and Pb. System and manufacturing costs are considerable.

Many transportation and non-planar surface waste heat recovery applications require high specific power and flexibility/conformability, making inorganic TE materials heavy and brittle. Non-toxic, solution-processable, low-density, mechanically flexible, thermally inefficient organic materials are cheap. Doping conductive and semiconducting organic materials chemically and electrochemically changes their electronic structure. Electrical conductivity is good with PEDOT and poly(styrenesulfonate). Electrical conductivity is enhanced by solvents, salts, zwitterions, carboxylic or inorganic acids, polar organic molecules, and cosolvents. Increasing oxidation levels enhances carrier mobility, Seebeck coefficient, and electrical conductivity through dedoping and doping processes.

Traditional TE devices are made by ball milling or melting/heating components. The alloy powders are consolidated and diced. Insulating substrate electrical shunts interface with TE “legs”. Classic TE leg fabrication chips or microcracks fragile materials, limiting size, shape, and packing density. Thin film TE materials are made from supralattices and nanowires in semiconductor manufacturing. Solution-based printing is affordable and easy for large-scale fabrication. Commercial exhibits display printed TE devices. Screen printing, inkjet, direct write, and stereolithography deposit TE.

Scalability involves making many devices quickly, with low waste, and changing design or scale. Material and device advantages are contrasted. These examples calculate ZT using power factor and printed material element thermal conductivity weighted averages [22].

3.5 CHARACTERIZATION AND TESTING OF THERMOELECTRIC DEVICES

Performance and current–voltage ratings rely on the module's cross-sectional area, thickness, and particle count [23]. Testing began with the thermoelectric converter's V/I output (Figure 3.10). All linear V/I graphs have nearly identical slopes. The thermoelectric converter's internal resistance is maintained during load and test temperature operations.

An Indium Tin Oxide/Platinum (ITO/Pt) thermocouple's thermoelectric response in 0.15 Pa O_2, 0.4 Pa N_2, and 1.33 Pa argon is shown in Figure 3.11. Figure 3.11 exhibits 176°C gradient-induced mild inflections at 920°C. The oxygen partial pressure made high-temperature ITO thin films p-type. Figure 3.12 shows the ITO/Pt thermocouples' heat cycling thermoelectric response. ITO-argon thermoelements are shown. The dark grey curve shows a 63 mV emf at a 1298°C hot junction temperature.

When heat gradients increased and decreased, Type S thermocouples had much lower output voltages. Large heat gradients masked behavior, lowering divergence. Thermocouple Seebeck coefficient-based thermoelectricity was measured using a third-order polynomial.

Figures 3.13 and 3.14 indicate that test selection continuously assesses device performance. In TE system evaluation, essential parameters such as P_{max}, W/g, Ri, P−I, and η are measured using the device, open circuit, and current. A rapid rise in material resistance (V_d/V_{oc} reduction) suggests defective legs/devices [26–30].

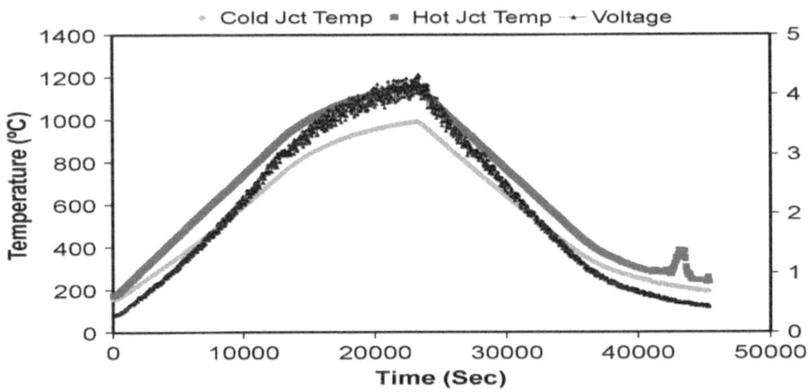

FIGURE 3.11 Comparison of ITO thermoelement's thermoelectric response to Pt in an environment with 0.15 Pa O_2, 0.40 Pa N_2, and 1.33 Pa Ar [25].

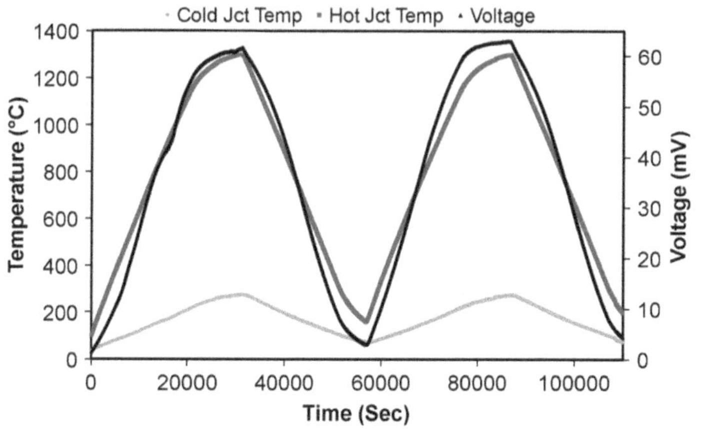

FIGURE 3.12 Relative thermoelectricity of ITO and platinum [25].

3.6 ADVANCED MANUFACTURING APPROACHES AND EMERGING TECHNOLOGIES

Processing and assembling are typical in thermoelectric device production. Mechanical thermoelectric compound alloying, hot pressing, or plasma sintering of alloy powders are also involved. Additive manufacturing could help thermoelectric devices. Component assembly elimination could minimize the complex thermoelectric module production time and cost.

3.6.1 Additive Manufacturing (AM)

Personalized biomedical implants, soft robotics, and wearable electronics have popularized soft AM. Soft devices are evolving fast; thus, one-size-fits-all may be outdated. As traditional manufacturing becomes more expensive and harder to create novel geometries, soft material processing is moving to AM. AM creates layers of computer-designed structures via nozzle deposition, binder jetting, photopolymerization, or thermal coalescence. AM technologies offer complex geometries, free customization, and geometric design. Advanced geometry softens stiff materials.

Polymers and polymer composites can deform without breaking due to lattice and auxetic features. Many biological uses require body-shaped material. The use of multiple-material layers, intelligent structure design, stimulus-responsive 3D geometries, and functionality gradients have created functional materials in several AM technologies [31].

Cryogenically crushed ABS polymer pellets into 400 μm powder for homogeneous mixing. Sample production began with mixing Bi_2Te_3 and ABS powder in a Filastruder extruder at 220 °C. The Ultimaker 2 used stock custom filament [32].

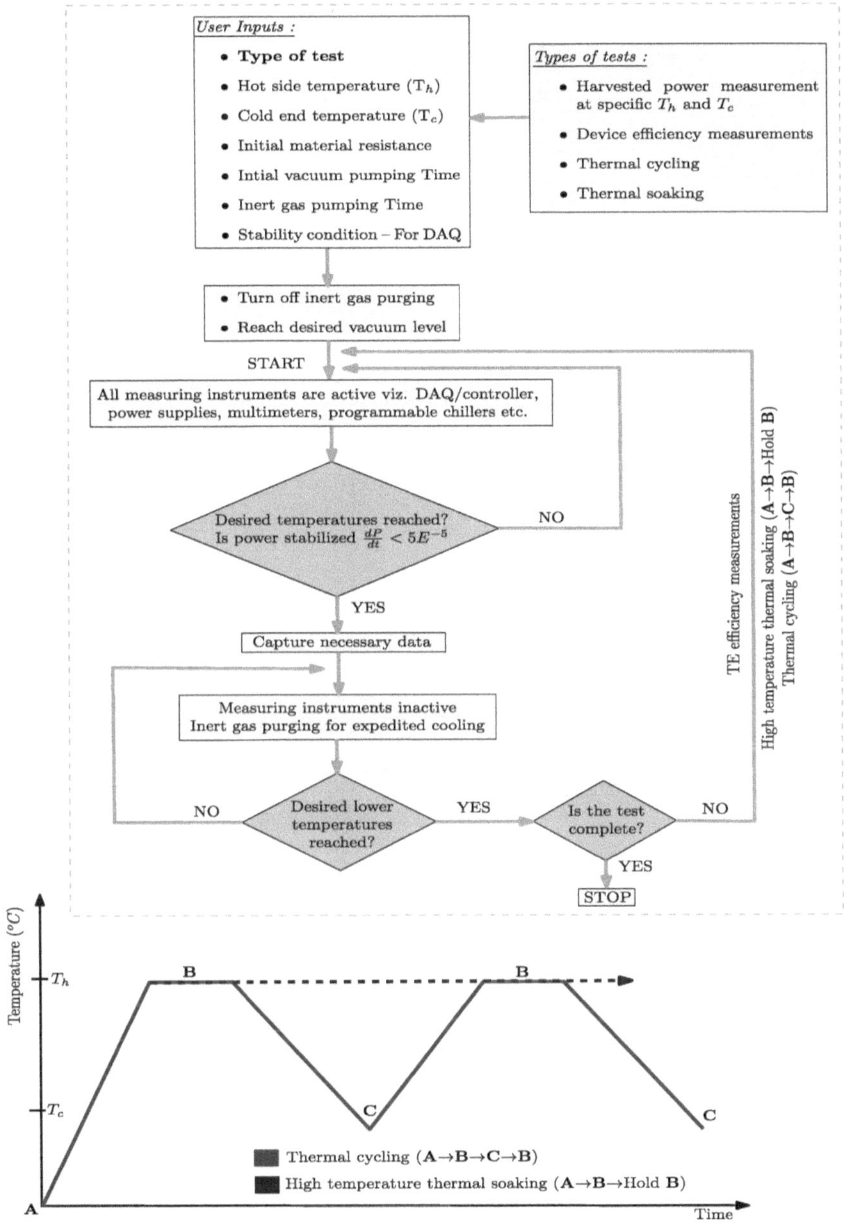

FIGURE 3.13 LabVIEW® software tests harvested power, device efficiency, thermal cycling, and high-temperature soaking at T_c and T_h [26].

All blends were 80% Bi_2Te_3 to ABS, since higher ratios produced brittle filaments that broke in the 3D printer tube. At 260°C, maintain a 0.4mm nozzle temperature. For material deposition, the heat bed was heated to 90°C.

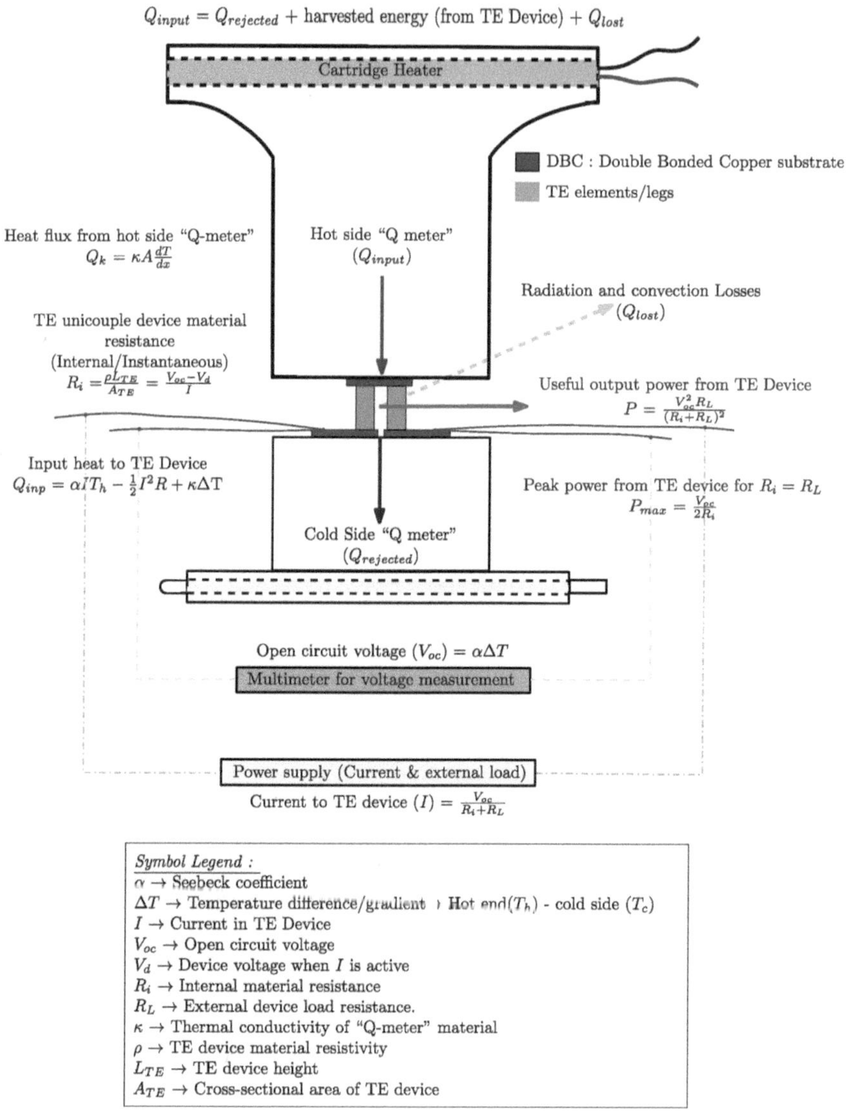

FIGURE 3.14 TE sample between hot (T_h) and cold (T_c) ends [26].

In this investigation, bulk Bi_2Te_3 has twice the ZT of additively generated samples. A SEM image shows that printed samples' porosity influences electrical conductivity. Band engineering, nano-engineering, thermoelectric doping, and mechanical strengthening for bigger geometries will be pursued. Using selective laser melting (SLM), many researchers have created intricate stainless steel, titanium, and nickel superalloy parts. Testing Bi_2Te_3 powder compact single melt tracks with a low repetition pulsed laser. Only thermoelectrically pre-alloyed

Bi_2Te_3 is sold. Off-the-shelf Bi_2Te_3 powder cannot print identical SLM thin layers. Powder compacting laser-irradiated Bi_2Te_3 investigates how laser energy density affects microstructural evolution and densification. These results enable layer-by-layer SLM thermoelectric module processing [33].

3.6.2 Advanced Manufacturing Approaches and Emerging Technologies

Semimetals and semiconductors with added impurities are the primary focus of study in the field of thermoelectrics. The most recent thermoelectric (TE) materials include chalcogenides, silicides, clathrates, skutterudites, half-Heusler alloys, and oxides, which show potential advancements in tetrahedrite. Bismuth telluride TE devices are commonly used for localized heating and cooling in commercial applications.

The introduction of commercial TE devices for large-scale waste heat recovery and industrial process prototypes and feasibility studies was recent. Scarcity, toxicity, high raw material costs, and the processing of Bi, Te, Sb, and Pb make TE devices challenging to commercialize. Manufacturing and system costs are high. Many transportation and non-planar surface waste heat recovery applications require high specific power and flexibility/conformability, but inorganic TE materials are too heavy and brittle.

Industrial process prototypes and feasibility studies are available for new commercial TE devices for large-scale waste heat recovery. TE devices are expensive to manufacture and systemize, and scarcity, toxicity, high raw material costs, and the processing of Bi, Te, Sb, and Pb preclude widespread commercialization. Inorganic TE materials are too heavy and brittle for transportation and non-planar surface waste heat recovery applications that require high specific power and flexibility/conformability.

Electrical conductivity rises with organic solvents, salts, zwitterions, carboxylic or inorganic acids, polar organic molecules, or cosolvents. Oxidation increases the Seebeck coefficient, electrical conductivity, and carrier mobility in doping and dedoping. Samples of thermal, electrical, and Seebeck coefficients were handled differently.

The estimated ZT has a large uncertainty range due to 3×x method uncertainties in extracting in-plane thermal conductivity. All TE properties should be characterized in the same direction because TE materials carry heat and electricity equally. The in-plane thermal conductivity of spin-coated PEDOT:PSS is higher than through-plane due to electrical conductivity anisotropy and substantial electron contribution. Recently, PEDOT:PSS thin films showed higher in-plane thermal conductivity than time-domain thermoreflectance, flash analysis, and suspended microdevice results. Comparisons of study materials should be avoided due to processing variations.

The geometry of PEDOT:PSS films affects their electrical conductivity and TE performance. PEDOT:PSS phase-segregated films have PSS shells around grains. PEDOT-depleted grain borders, which are less electrically conductive than grain cores, limit PEDOT:PSS charge transmission. PEDOT:PSS's pancake-shaped PEDOT-rich grain islands form an electrical percolation network.

Grain orientation changes during deposition. Electrical conductivity anisotropy in spin-coated samples can be three orders of magnitude because grains align parallel to the film plane. Inkjet-printed films exhibit the coffee ring effect because evaporation-driven internal convective fluxes orient granules both parallel and perpendicular to the film plane. Electrical conductivity is 1.3:1.17 out-of-plane to in-plane due to orientation.

Positive (bi) polarons travel between sites in conductive polymers at ambient temperature via delocalized diffusion or phonon-assisted, localized hopping in response to atom vibrations. Molecular shape, crystallinity, doping, stacking, and conformation affect both transport modalities. Carrier concentrations, monomer and counterion chemical structures, and molecular organization affect the TE properties of conjugated polymers. Not as much as polymers, charge transfer complexes, radical cation salts, and molecular semiconductors like pentacene and fullerene films, which are being studied for their TE characteristics 22].

TEGs can be made using several technologies, production processes, and substrate materials: Silicon technology produces micro- to nano-meter generators (Figure 3.15). TC arms are built from compatible (N- and P-type polycrystalline silicon, poly-SiGe, and subsequently SiNW) and incompatible CMOS materials like the Bi–Sb–Te alloy. CMOS technology creates micro-cavities in silicon substrates to reduce heat dissipation under hot junctions (Figure 3.15).

Alumina and low-temperature co-fired ceramic (LTCC) can be used to make high-density TEGs. TEG manufacturing is appropriate for ceramics due to their multilayered deposition, simplicity, low cost, time efficiency, and durability to high temperatures, gases, and stresses. Thin, thick, and mixed deposition can be

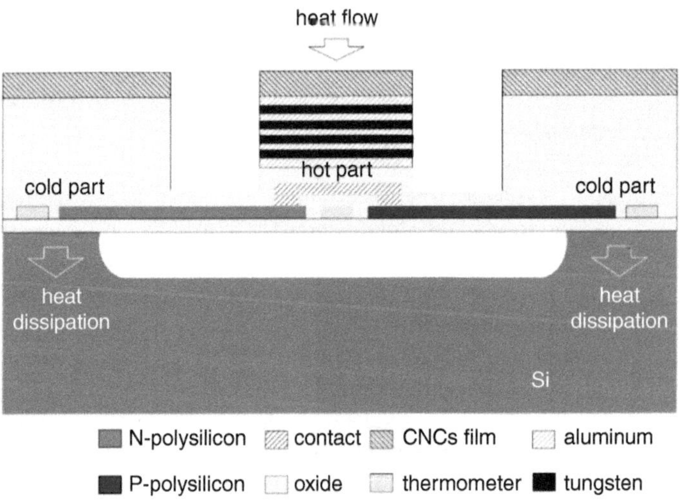

FIGURE 3.15 A mixed TEG with N- and P-type polysilicon thermocouples made with CMOS technology has a suspended hot portion [34].

used to make ceramic thermocouples. PdAg, Ni, Pt, Al, W, and Pd pastes are used in screen-printed thick-film thermocouples. Thin films are made via microelectronic deposition of Germanium-based compositions and group III-Nitride alloys like AlN and GaN. The Seebeck coefficient and electrical resistance are high in thin-film ceramic semiconductors.

The low thermal conductivity of ceramic substrates (3.3 W/K·m for DP951 green tape) causes the generator's thermal resistance. As a result, the LTCC-based TEG has a large output voltage and temperature gradient. To limit heat transfer between TC arms and maximize the hot-cold junction temperature difference, ceramic substrates are preferable for lateral TEGs (Figure 3.16a). The substrate can be punched with multiple holes filled with conductive paste to arrange the TC arms vertically for vertical TEGs (Figure 3.16b).

Flexible TEGs (f-TEGs) for wearable applications are best made from polymers, polyimide, cellulose fibers, and fabric substrates. Electro-deposition, sputtering, or evaporation are used for thin films, and screen-printing is used for thick films (Figure 3.17).

Many techniques have been tried to improve efficiency, but designing and producing new TE materials with appropriate Figure of Merit (zT) and Power Factor values is still difficult. The thermoelectric properties of n- and p-type materials determine a thermoelectric module's operation. Both n- and p-type legs must have equal Figures of Merit for high performance, although some compatible materials are too difficult for next-generation TEG model creation [35].

The high figure of merit has led to the production of SnSe by solvothermal synthesis or melting (M), spark plasma sintering (SPS), cold pressing, hot pressing, and thermal evaporation. Several methods demand high temperatures, pressure, and manufacturing time. Instead, printing can be done quickly and easily at room temperature and air pressure. Printing has been restricted to generating films with a thickness of 300 μm, except for bulk Bi_2Te_3 samples with a ZT of 0.9 at 398 K. The application of these coatings does not provide any advantages over traditional bulk thermoelectric generator legs, which have dimensions of 1 mm×1 mm×2mm. Contrary to Bi_2Te_3, there has been limited research conducted on printed thermoelectric materials at medium and high temperatures. There have been no reports of non-Bi_2Te_3 samples being printed in large quantities [36].

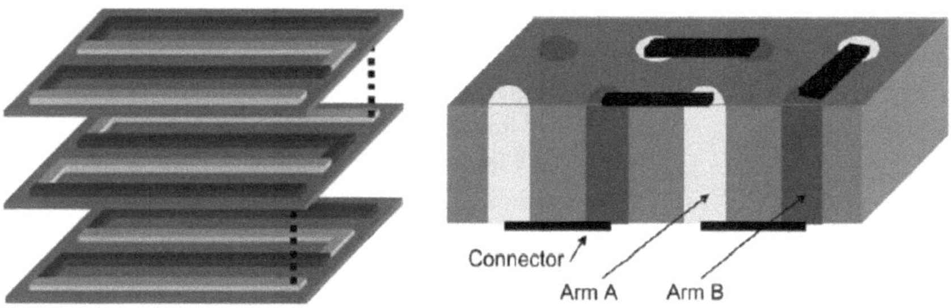

(a) (b)

FIGURE 3.16 LTCC-based multilayered TEG: (a) planar design using Ag/Ni thermocouples and (b) vertical design using conductive 34].

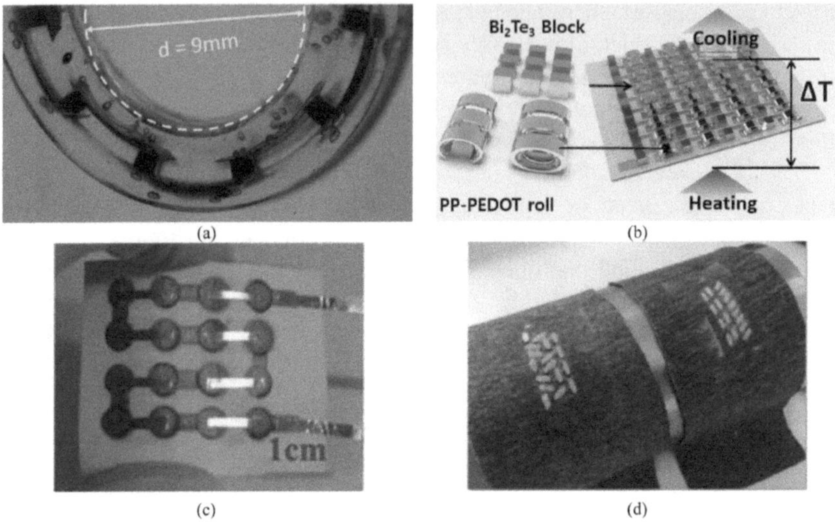

FIGURE 3.17 Vertical flexible TEGs fabricated on a substrate made of: (a) PDMS polymer, (b) polyimide (PI), (c) cellulose fiber (paper substrate with cooper/nickel TCs), and (d) knitted fabric [34].

3.6.3 TEG Materials

Thermoelectric (TE) modules have parallel n- and p-type semiconducting legs. The structure and content classification of thermoelectric materials include major classes such as Clathrate, Chalcogenide, Half-Heusler, Skutterudite, Silicide, and Oxide. Application issues include material selection and temperature-dependent thermoelectric properties. The figure-of-merit, ZT, shows the functionality of thermoelectric materials.

Thermoelectric modules have multiple n- and p-type couplings. Electrical insulators, usually ceramic, cover these pairs and connection points. Higher Seebeck coefficients and lower electrical and thermal resistance improve thermoelectric conversion efficiency. Three temperature-dependent thermoelectric approaches use heavily doped semiconductors.

Bismuth, Lead, and Silicon-Germanium alloys are commonly used in thermoelectrics. Optimizing the power factor requires adjusting the Seebeck coefficient and electrical conductivity with reduced Fermi energy. Materials composed of bismuth telluride have the highest figure-of-merit (3.4×10^{-3} K^{-1}) but are limited to temperatures below 250°C. Lead telluride has the second-highest mean figure-of-merit (2.0×10^{-3} K^{-1}) at temperatures up to 500°C.

While silicon-germanium has the lowest figure-of-merit (0.8×10^{-3} K^{-1}), it can function at temperatures around 1000°C for extended periods. Complex structures like Clathrates, Skutterudites, and Zintl phases improve ZT. New prospects have emerged from nanostructures and substructures. At room temperature, superlattice materials exhibit ZT up to 2.5 but are unsuitable for mass production. Nanostructures reach record thermal efficiency. Recent understanding of phonon scattering and energy filtering makes nanocomposites viable ZT materials. High-copper tellurium-based glasses validate chalcogenide semiconducting glasses for high-performance thermoelectric materials.

3.6.4 Thermoelectric Material Preparation

The n-type β-$FeSi_2$/Bi_2Te_3 graded thermoelectric material is made by dip coating with $Sn_{95}Ag_5$ as the bridge material. The maximum power output is 2.5/3 times that of monolithic material β-$FeSi_2$ at the same temperature difference. A complex sol-gel process is used to create polycrystalline $La_{1-x}Sr_xCuO_{3-\delta}$ samples. By replacing La atoms with strontium, the transit properties of the samples change.

Silicon molding produces 300-m-high, 40-m-wide Bi–Sb–Te thermoelectric components. Cylindrical explosive consolidation produces Bi–Te–Se and Bi–Te–Sb three-phase powders. In liquid nitrogen, cryogenic grinding produces 70-nm Bi_2Te_3 nanoparticles. Cryogenic grinding produces sinterable Bi_2Te_3 particles.

The current TE priority is developing high ZT TE materials to meet theoretical predictions and lower production costs to make TE technology a more competitive alternative energy source with various practical uses.

Optimizing ZT involves manipulating electrical transport properties (S, σ) and κ using various methods. Researchers want to increase ZT over unity by

1. Finding new materials with high ZT, and
2. Combining theoretical and experimental studies to decouple S and σ, lowering κ_L toward the amorphous limit.

More diversified than Bi_2Te_3-, PbTe-, and SiGe-based materials, many good TE materials have been found and manufactured in the past two decades.

High-performance TE material mechanisms have been shown by extensive theoretical and experimental experiments. Many strategies make it difficult to increase every parameter without lowering efficiency; therefore, large efforts are still needed to bring the ZT near the theoretical limit. After fundamental studies, synergetic optimization should develop TE materials. Optimize electrical transport in materials by optimizing n and/or band engineering, and reduce κ through nanostructure, hierarchical, defect, porosity, or grain boundary engineering. Metallic vanadium dioxide, with unusually low κe at its metal-insulator transition, may have the ability to separate κe and σ.

TE materials' future hinges on such fascinating and in-depth research, and revolutionary advancements are expected. The growing alternative energy industry benefits ZT and TE material consumption. More efficient and reasonable development of TE materials and gadgets is needed to boost their practicality. With higher specific power, smaller size, and less weight, radioisotope thermal generators (RTGs) will power remote missions due to their durability and reliability. TEGs will be used more in automotive exhaust waste heat recovery and temperature conditioning to reduce emissions and fuel consumption.

MEMS and conductive polymers will enable highly integrated implantable/wearable TE devices for medical support, biodata sensing and collection, body temperature monitoring and control, entertainment, and micromechanical cooling. Investigate prospective micro-TE device applications that transform other signals into heat change. High light-confinement, extremely wavelength-dependent TE nano-photodetectors manage nanoscale volume temperature gradients and give a non-band gap-limited, uncooled, filterless two-material spectrometer from visible to infrared on one chip. Small size makes this nano-photodetector's rise time 100 times faster than bulk TE detectors, enabling new TE nanodevices [37].

REFERENCES

1. Soprani, S., Haertel, J. H. K., Lazarov, B. S., Sigmund, O., & Engelbrecht, K. (2016). A design approach for integrating thermoelectric devices using topology optimization. *Applied Energy*, *176*, 49–64.
2. He, R., Schierning, G., & Nielsch, K. (2018). Thermoelectric devices: a review of devices, architectures, and contact optimization. *Advanced Materials Technologies*, *3*(4), 1700256.
3. Zhao, D., & Tan, G. (2014). A review of thermoelectric cooling: materials, modeling and applications. *Applied Thermal Engineering*, *66*(1–2), 15–24.
4. LeBlanc, S., Yee, S. K., Scullin, M. L., Dames, C., & Goodson, K. E. (2014). Material and manufacturing cost considerations for thermoelectrics. *Renewable and Sustainable Energy Reviews*, *32*, 313–327.
5. El-Desouky, A., Carter, M., Andre, M. A., Bardet, P. M., & LeBlanc, S. (2016). Rapid processing and assembly of semiconductor thermoelectric materials for energy conversion devices. *Materials Letters*, *185*, 598–602.
6. Zhang, T. (2016). Design and optimization considerations for thermoelectric devices. *Energy Conversion and Management*, *112*, 404–412.
7. Stevens, J. W. (1999, August). Heat transfer and thermoelectric design considerations for a ground-source thermo generator. In *Eighteenth International Conference on Thermoelectrics. Proceedings, ICT'99*, Baltimore, MD (Cat. No. 99TH8407) (pp. 68–71). IEEE.
8. Schierning, G., Chavez, R., Schmechel, R., Balke, B., Rogl, G., & Rogl, P. (2015). Concepts for medium-high to high temperature thermoelectric heat-to-electricity conversion: a review of selected materials and basic considerations of module design. *Translational Materials Research*, *2*(2), 025001.
9. Sarma, D. D. (2021). Essential considerations for reporting thermoelectric properties. *ACS Energy Letters*, *6*(10), 3715–3718.
10. Wang, Y., Lin, P., Lou, Q., Zhang, Z., Huang, S., Lu, Y., & He, J. (2021). Design guidelines for chalcogenide-based flexible thermoelectric materials. *Materials Advances*, *2*(8), 2584–2593.
11. Lara Ramos, D. A., Barati, V., Garcia, J., Reith, H., Li, G., Pérez, N., & Nielsch, K. (2019). Design guidelines for micro-thermoelectric devices by finite element analysis. *Advanced Sustainable Systems*, *3*(2), 1800093.
12. Kiziroglou, M. E., Wright, S. W., Toh, T. T., Mitcheson, P. D., Becker, T., & Yeatman, E. M. (2013). Design and fabrication of heat storage thermoelectric harvesting devices. *IEEE Transactions on Industrial Electronics*, *61*(1), 302–309.
13. Konstantinou, G., Kyratsi, T., & Louca, L. S. (2022). Design of a thermoelectric device for power generation through waste heat recovery from marine internal combustion engines. *Energies*, *15*(11), 4075.

14. Jouhara, H., Żabnieńska-Góra, A., Khordehgah, N., Doraghi, Q., Ahmad, L., Norman, L., ... & Dai, S. (2021). Thermoelectric generator (TEG) technologies and applications. *International Journal of Thermofluids, 9*, 100063.
15. Luo, D., Yan, Y., Chen, W. H., Yang, X., Chen, H., Cao, B., & Zhao, Y. (2023). A comprehensive hybrid transient CFD-thermal resistance model for automobile thermoelectric generators. *International Journal of Heat and Mass Transfer, 211*, 124203.
16. Kanimba, E., Pearson, M., Sharp, J., Stokes, D., Priya, S., & Tian, Z. (2018). A comprehensive model of a lead telluride thermoelectric generator. *Energy, 142*, 813–821.
17. Huang, G. Y., Hsu, C. T., Fang, C. J., & Yao, D. J. (2016). Optimization of a waste heat recovery system with thermoelectric generators by three-dimensional thermal resistance analysis. *Energy Conversion and Management, 126*, 581–594.
18. Jia, X. D., Wang, Y. J., & Gao, Y. W. (2017). Numerical simulation of thermoelectric performance of linear-shaped thermoelectric generators under transient heat supply. *Energy, 130*, 276–285.
19. Luo, D., Yan, Y., Li, Y., Wang, R., Cheng, S., Yang, X., & Ji, D. (2023). A hybrid transient CFD-thermoelectric numerical model for automobile thermoelectric generator systems. *Applied Energy, 332*, 120502.
20. Luo, D., Liu, Z., Yan, Y., Li, Y., Wang, R., Zhang, L., & Yang, X. (2022). Recent advances in modeling and simulation of thermoelectric power generation. *Energy Conversion and Management, 273*, 116389.
21. Luo, D., Yan, Y., Wang, R., & Zhou, W. (2021). Numerical investigation on the dynamic response characteristics of a thermoelectric generator module under transient temperature excitations. *Renewable Energy, 170*, 811–823.
22. Orrill, M., & LeBlanc, S. (2017). Printed thermoelectric materials and devices: fabrication techniques, advantages, and challenges. *Journal of Applied Polymer Science, 134*(3), 1–15.
23. Montecucco, A., Buckle, J., Siviter, J., & Knox, A. R. (2013). A new test rig for accurate nonparametric measurement and characterization of thermoelectric generators. *Journal of Electronic Materials, 42*, 1966–1973.
24. Carmo, J. P., Antunes, J., Silva, M. F., Ribeiro, J. F., Goncalves, L. M., & Correia, J. H. (2011). Characterization of thermoelectric generators by measuring the load-dependence behavior. *Measurement, 44*(10), 2194–2199.
25. Gregory, O. J., Busch, E., Fralick, G. C., & Chen, X. (2010). Preparation and characterization of ceramic thin film thermocouples. *Thin Solid Films, 518*(21), 6093–6098.
26. Rao, A., Joshi, G., Poudel, B., & Ballinger, C. (2022). A custom designed modular, scalable test system for an efficient performance evaluation of thermoelectric devices. *Energy Conversion and Management: X, 14*, 100228.
27. Rao, A., Bosak, G., Joshi, B., Keane, J., Nally, L., Peng, A., … & Poudel, B. (2017). A tialcu metallization for 'n' type cosbx skutterudites with improved performance for high temperature energy harvesting applications. *Journal of Electronic Materials, 46*(4), 2419–2431
28. Rowe, D. M. (2005). *Thermoelectrics Handbook: Macro to Nano.* CRC Press.
29. Lau, P. G., & Buist, R. J. (1997). Calculation of thermoelectric power generation performance using finite element analysis. In *Proceedings ICT'97. XVI International Conference on Thermoelectrics* (pp. 563–566). IEEE, Dresdan, Germany.
30. Rao, A., Banjade, P., Bosak, G., Joshi, B., Keane, J., Nally, … &, Poudel, B. (2016) A quick and efficient measurement technique for performance evaluation of thermoelectric materials. *Measurement Science and Technology, 27*(10), Article 105008.

31. Regis, J. E., Renteria, A., Hall, S. E., Hassan, M. S., Marquez, C., & Lin, Y. (2021). Recent trends and innovation in additive manufacturing of soft functional materials. *Materials*, *14*(16), 4521.
32. Oztan, C., Ballikaya, S., Ozgun, U., Karkkainen, R., & Celik, E. (2019). Additive manufacturing of thermoelectric materials via fused filament fabrication. *Applied Materials Today*, *15*, 77–82.
33. El-Desouky, A., Carter, M., Mahmoudi, M., Elwany, A., & LeBlanc, S. (2017). Influences of energy density on microstructure and consolidation of selective laser melted bismuth telluride thermoelectric powder. *Journal of Manufacturing Processes*, *25*, 411–417.
34 Jaziri, N., Boughamoura, A., Müller, J., Mezghani, B., Tounsi, F., & Ismail, M. (2020). A comprehensive review of thermoelectric generators: technologies and common applications. *Energy Reports*, *6*, 264–287.
35. Jouhara, H., Żabnieńska-Góra, A., Khordehgah, N., Doraghi, Q., Ahmad, L., Norman, L., & Dai, S. (2021). Thermoelectric generator (TEG) technologies and applications. *International Journal of Thermofluids*, *9*, 100063.
36. Burton, M. R., Mehraban, S., Beynon, D., McGettrick, J., Watson, T., Lavery, N. P., & Carnie, M. J. (2019). 3D printed SnSe thermoelectric generators with high figure of merit. *Advanced Energy Materials*, *9*(26), 1900201.
37. Yang, L., Chen, Z. G., Dargusch, M. S., & Zou, J. (2018). High performance thermoelectric materials: progress and their applications. *Advanced Energy Materials*, *8*(6), 1701797.

4 Thermoelectric Material Selection and Characterization for Heating and Cooling Applications

4.1 INTRODUCTION

The most popular thermoelectric materials, Bi_2Te_3 and PbTe alloys, have ZT values near unity (at ambient temperature for PbTe and 500–700 K for Bi_2Te_3). High electrical and low heat conductivity are the main required properties of thermoelectric materials. Thermoelectric materials have been studied to increase the Seebeck coefficient and decrease heat conductivity by changing their nanostructure. High electrical conductivity materials require unique methods that can be applied at 450 K due to the relationship between charge carriers and thermal and electrical conductivity. Lead (Pb) alloys make materials that can survive 850 K, while silicon germanium (SiGe) alloys make 1300 K thermal elements.

Low-temperature materials are made from bismuth (Bi) alloys with tellurium (Te), antimony (Sb), and selenium (Se). These alloys can be used at 450 K. SiGe alloys may operate up to 1300 K, while lead (Pb) alloys work up to 850 K. Despite developments in thermoelectric power production due to new materials and material architectures with higher thermoelectric performance, materials remain vital for practical and commercial applications. Researchers have reduced lattice thermal conductivity to improve the material's figure of merit and conversion efficiency.

The semiconductor compound ß-Zn_4Sb_3 performs well in the ZT temperature spectrum between Bi_2Te_3 and PbTe. Thermal conductivity is quite low, and its maximal ZT is 1.3 at 670 K. This cheap, vacuum-stable material is stable at this temperature. By raising the figure of merit, researchers try to make thermoelectric materials more competitive [1].

TE devices react differently to power factor (PF) and thermal conductivity. Explaining this issue requires examining material qualities and device performance.

DOI: 10.1201/9781003639343-4

The effects of material quality on system design reveal that filler, substrates, solders, contacts, manufacturing technique, and cost all play a role in TE device design. Material, chemical, and thermal stability, interface engineering, and hot and cold side thermal resistance optimization are key to marketing TE systems. The majority of functional materials in commercial TE modules are telluride and bismuth. Financial, ecological, and recyclable issues aside, the gap between TE materials' actual and expected high-efficiency values hinders their broad implementation. Traditional TE device efficiency has been compared to energy conversion system design to optimize ZT, optimizing power output in specific purposes like small TE generators [2].

Materials with significant promise have carrier densities of 10^{19} carriers/cm^3. Doping often raises the power factor (0–20), which raises zT. The highest mobility carriers give the maximum electrical conductivity for a given carrier concentration. The lattice thermal conductivity is being reduced without lowering the power factor to improve the materials' figure of merit (Table 4.1) [3].

TABLE 4.1
Properties of Interest for Thermoelectric Materials [3]

Sample Property	Relationship
Seebeck coefficient	$\alpha = \frac{-\Delta V}{\Delta T}$
Electrical resistivity	$\rho = R_s \times \frac{A}{l_0}$
Electrical conductivity	$\sigma = ne\,\mu = \frac{1}{\rho}$
Thermal conductivity	$\lambda = \frac{Q_k l_0}{A\Delta T}$
Hall voltage	$V_H = \frac{R_H BI}{t}$
Hall coefficient (single-cartier system)	$R_H = \frac{1}{ne}$
Cartier mobility	$\mu_H = \sigma R_H$
Cartier concentration	$n = \frac{1}{R_H e}$

4.2 OVERVIEW OF HEATING AND COOLING REQUIREMENTS

4.2.1 Heating Requirements

Nanostructuring to modify electron and phonon transport has greatly improved the efficiency of typical inorganic crystalline thermoelectric materials over the past 20 years. Despite these advances, Bi_2Te_3 and its alloys dominate near-ambient applications. However, these improvements largely apply to high-temperature applications. The room-temperature stability of various organic semiconducting materials (photoinduced oxidation) makes them useful in many applications. In addition, power conversion efficiency changes as $\Delta T/T_{avg}$, or temperature differential divided by temperature average, making expensive present thermoelectric modules unfeasible for harvesting thermal energy near ambient temperature. Localized cooling and heating are the main commercial uses of room-temperature thermoelectrics.

The carrier concentration is notoriously difficult to decouple from the thermoelectric properties. Electrical conductivity is impeded by morphological and electronic flaws, but the density of states directly affects the Seebeck coefficient. The Seebeck coefficient, which measures charge carrier average entropy, decreases with carrier concentration [4].

Thin films of various TE materials are developed from 5 to 30 mm. These materials are manufactured into submillimeter-pair arrays using standard semiconductor fabrication processes. The devices may heat, cool, or create energy with thermal reaction times of 20 ms or less [5].

Many low ZT materials have lower lattice thermal conductivities than others. Lattice thermal conductivity reduction improves TE efficiency, ZT, and carrier concentration re-optimization, improving ZT even more. By alloying Bi_2Te_3 with Sb_2Te_3 and Bi_2Se_3, the carrier concentration can be accurately controlled while maintaining lattice thermal conductivity.

Many studies focus on p-type compositions $(Sb_{0.8}Bi_{0.2})_2Te_3$, but few on n-type compositions. Waste heat recovery is being studied using Bi_2Te_3 thermoelectric modules. The data demonstrate that output power and current increase with temperature difference. Most power generation at moderate temperatures (500–900 K) uses group-IV tellurides like PbTe, GeTe, or SnTe. Several alloys, including $AgSbTe_2$, have caused ZT in n- and p-type materials. High-temperature TEGs (4900 K) have n- and p-type legs made of silicon-germanium alloys.

Due to diamond's great lattice thermal conductivity, the ZT is low for p-type materials. Studies show zinc antimony (Zn_4Sb_3) is a powerful TE material. Its high efficiency comes from its semiconductor-like electrical structure and heavy doping. Due to diamond's great lattice thermal conductivity, the ZT is low for p-type materials. A highly effective TE material is zinc antimony (Zn_4Sb_3).

Because of its limited heat conductivity, its severely doped semiconductor-like electrical structure maximizes efficiency. Figure 4.1 shows a generator with TEG modules. To take advantage of the generator's continuous temperature differential between the front and back ends, TEGs of varying performance capacities

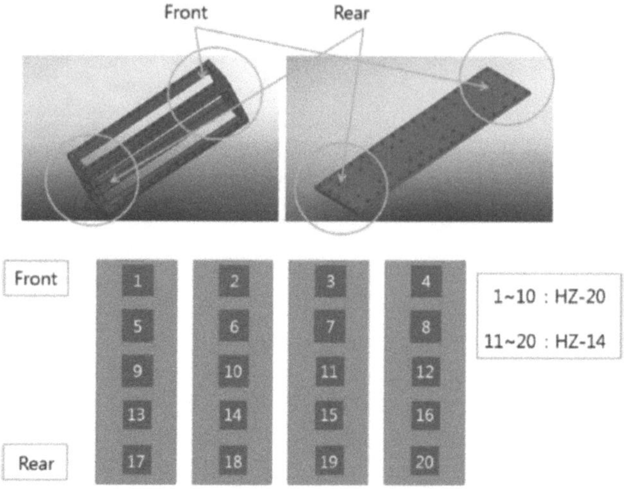

FIGURE 4.1 The thermoelectric generator's schematic design for each module placement [6].

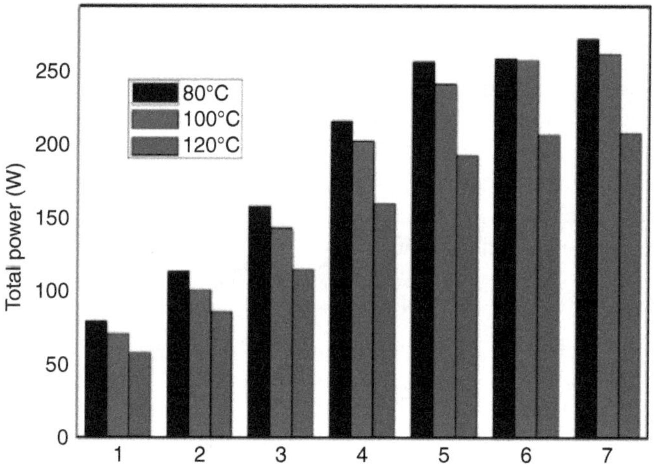

FIGURE 4.2 Power output comparison across various temperatures [6].

are strategically placed. The HZ-14's limited range and poor performance in high weather require modules to be moved to the cooler back of the vehicle. Front-mounted HZ-20 modules have a larger temperature range. Power generation increases with the TEG hot-cold temperature difference. Power production decreases with temperature, as shown in Figure 4.2. The heat source or heating temperature affects performance more than the heat sink's flow pattern or water flow velocity. Thus, to improve TEM performance, reduce water flow rate to reduce power consumption and increase hot-side temperature to increase temperature differential [6].

4.2.2 Cooling Requirements

In 1834, the Peltier effect solid-state active heat pumps were discovered as thermoelectric coolers (TECs). The Peltier cooling TEC circuit transfers heat between hot and cold sides using direct current. Because of their light weight, compact size, and lack of working fluid or mechanical moving components, basic TEC architectures can provide miniaturized and lightweight refrigeration equipment. Current affects Peltier power. Because of its benefits, thermoelectric (TE) cooling has been popular since the 1950s.

Complex high-efficiency material design insights are also helpful. The average ZT value and the possibility of replacing high-greenhouse gas (GHG) cooling agents with thermoelectric systems determine thermoelectric materials' power generation and household usage [7].

Despite 20 years of discovery, thermoelectric cooling materials with zT > 1 at medium and high temperatures have not progressed enough. At ambient temperature, only Bi_2Te_3 alloys, $CsBi_4Te_6$, and $Mg_3Bi_{2-x}Sb_x$ perform well. Below 150 K, only $Bi_{1-x}Sb_x$ performs well. High zT is impossible at low temperatures due to low T. A material with a high z must have contradicting properties, which is difficult. Mostly, thermoelectric cooling performance is studied for $Bi_{1-x}Sb_x$, Bi_2Te_3 alloys, and $Mg_3Bi_{2-x}Sb_x$. Low conduction-valence band overlap energies characterize Bi and antimony semimetals.

Thermoelectric performance is improved by the high Seebeck coefficient and poor heat conductivity of semiconducting $Bi_{1-x}Sb_x$. The maximal z value below 150 K makes single-crystalline $Bi_{1-x}Sb_x$ ideal for thermoelectric cooling. The Ettingshausen effect occurs when a transverse magnetic field creates a mutually perpendicular electric current and temperature differential. Ettingshausen involves electrons and holes, unlike Peltier.

The Peltier effect was seen in freezing water droplets at bismuth-antimony junctions. Early Peltier cooling measurements employed $Bi_{91}Sb_9$ and $Bi_{95}Sn_5$ semimetals with a unicouple, yielding a ΔT of only 10 K. Bi_2Te_3 semiconductors could cool thermoelectrically in the 1950s. The p-type Bi_2Te_3 and n-type Bi unicouple had a ΔT_{max} of 26 K at 285 K hot side. Later, n-type Bi_2Te_3 was created, with a ΔT_{max} of 40 K for both n- and p-type unicouples. Zone-melted Bi_2Te_3 was cooler due to homogeneity, impurity control, and orientation. Recent advances include solid-solution alloying for lattice thermal conductivity reduction. This led to Bi_2Te_3 alloy unicouples with ΔT_{max} above 60 K. Large-scale manufacture of Bi_2Te_3 alloys with higher mechanical characteristics, utilizing a cost-effective technique with strict quality control, is as crucial as zT improvement for commercial applications. Bi_2Te_3 alloys are used for thermoelectric cooling, with n-type $Bi_2Te_{3-x}Se_x$ having a lower zT than p-type $Bi_{2-x}Sb_xTe_3$ [8].

Exfoliated graphene in solvents makes flexible TE films. Inorganic TE thin films on flexible surfaces increase flexibility. Industrial usage of intrinsic flexible Organic Thermoelectric (OTE) materials is possible due to their low toxicity, cost, and production. Doping and film crystallinity increase Peltier cooling electric transfer.

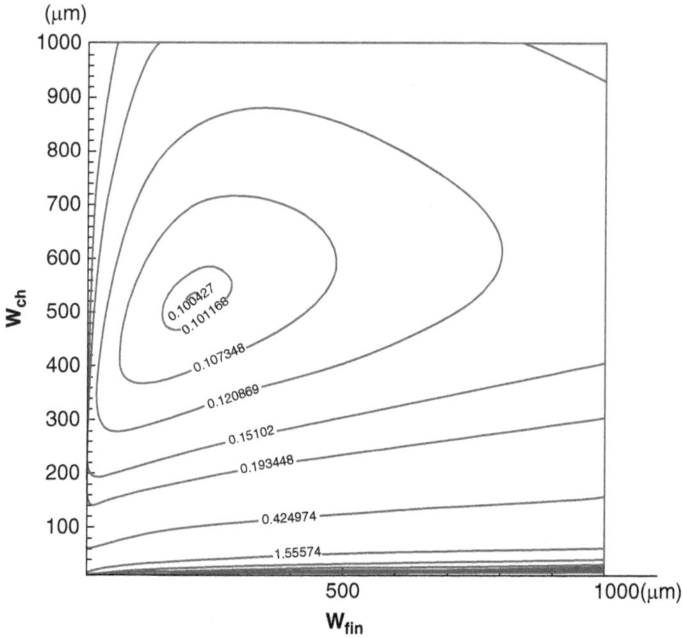

FIGURE 4.3 Microchannel heat sink air-cooled thermal resistance by fin and channel widths. 10 W pumping, 300 mm Lch [10].

Inorganic flexible device mechanics and physics encourage Peltier cooling. On soft polymeric or plastic substrates, inorganics are vapor-deposited or inkjet-printed. High tensile strength, low thermal conductivity, and mechanical performance make polyimide (PI) useful. Flexible devices with three p-n legs cooled steadily at 1.5 V [9]. Figures 4.3 and 4.4 show air- and water-cooled micro channel heat sink thermal resistances dependent on channel geometry.

Thermal conductivity lowers thermal resistance in fluid-cooled microchannel heat sinks. Water-cooling microchannel heat sinks are much smaller. Imposed regimes require liquid-cooled microchannel TEC heat sinks. Effective coolant pumps help microchannel heat sinks [10].

4.3 MATERIAL PROPERTIES FOR HEATING AND COOLING APPLICATIONS

4.3.1 Material Properties for Heating Applications

Silicon-based silicides are widespread due to their inexpensive cost and air temperature stability. These synthesized materials varied with upgraded qualities. Polymers, nanostructured bismuth telluride, MgAgSb, and polymer–inorganic matrices are low-temperature thermoelectric materials. Superior material properties increase high-temperature thermoelectric ZT values. Hierarchical

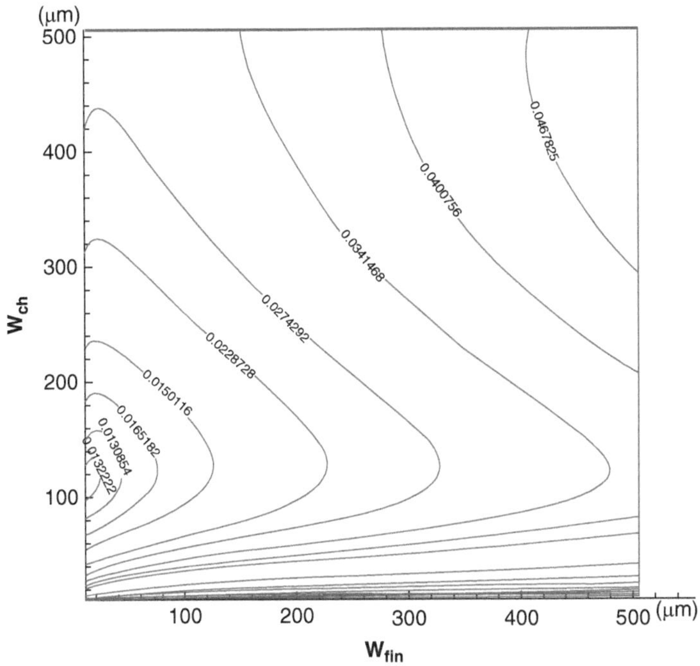

FIGURE 4.4 Channel and fin widths affect water-cooled microchannel heat sink thermal resistance. Pumping power=4 W, Lch=300 μm [10].

nano-to-meso-scale structure and tetrahedrites helped. High-temperature materials peak ZT above 600°C. Due to system component thermal resistances, the thermoelectric material hot side is lower than the heat source. Process furnaces and exhaust streams use heat sources over 600°C, and many 250°C–500°C applications generate thermoelectric power. Both applications' temperature range average ZT affects material selection and design. In Figure 4.5, high peak ZT materials and commercial and product development materials have ZT values.

Figure 4.5 indicates that high peak ZT values can mask the necessity for high application temperature average ZT values. At 642°C, SrTe-endotaxially nanostructured P-type PbTe exhibits 2.2 ZT. Schutterudite, tetrahedrite, and half-Heusler minerals have average ZT values that interest enterprises [11].

Peak ZT-rich N-type Mg_3Sb_2-TE materials predominate. Mid-temperature TE materials with high ZT based on Mg_3Sb_2 are cheap and toxin-free. Active devices/modules need low-resistance, strong-bonding TE contact layers. Sputtered thicker metallization layers on TE ends at ~200°C prevent electrode diffusion and lower contact resistance. Mass-producing such devices is expensive and weakens bonding. In series, TE modules use n- and p-type materials. TE materials with similar compounds and high compounds for near thermal expansion coefficients are needed for real-world applications. Similar compounds struggle in n- and p-TE.

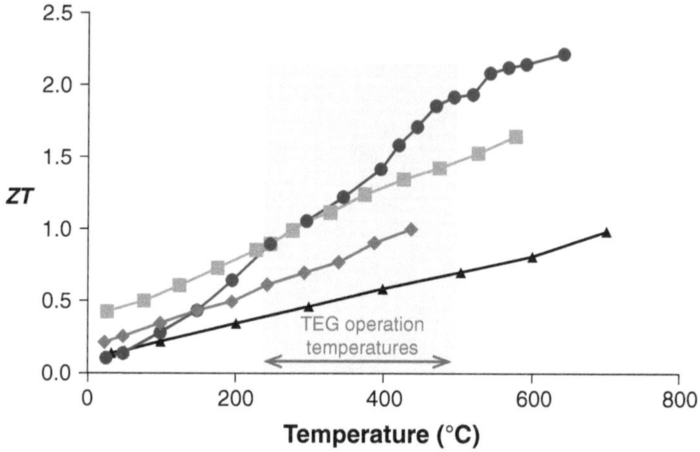

FIGURE 4.5 Overlaid on ZT versus temperature data are waste-heat source and TEG hot-side parameters [11].

A difference in coefficients causes thermal stress in p-type GeTe and n-type Mg_3Sb_2. Due to the series connection, all module TE legs should be well-attached to electrodes. TE legs are hard to make equal in length [12].

Potential Cu-based thermoelectric materials include diamond-like compounds, superionic conductors, tetrahedrites ($Cu_{12}Sb_4S_{13}$), and oxyselenides (BiCuSeO). Superionic Cu-based binary compounds have high thermal transport and electrical conductivity. Cu-based materials have ZTs > 2. In civil applications, their low cost and environmental friendliness are appealing. Minimizing Cu ion migration and creating n-type Cu-based thermoelectric materials are the major hurdles. For thermoelectric material reliability and lifespan, Cu ion migration must be understood and controlled. Management will yield various interesting and high-performance thermoelectric materials.

Despite lacking Te and poisonous Pb, CuSnSe and CuSnS exhibit good ZT. With k = 1 W/m·K at 300 K and a high Seebeck coefficient, $Cu_4Sn_7S_{16}$ may be an n-type Cu-based thermoelectric material. SnS polycrystals, a common and environmentally friendly thermoelectric material, have ZT greater than 1. Despite having the same crystal structure as SnSe, Na-doped SnS has low ZT. High heat conductivity and low power factor degrade SnS performance. Ag- or Na-doped polycrystalline SnS improves energy efficiency, carrier mobility, and thermal conductivity. $Sn_{0\cdot99}Ag_{0\cdot005}S$ holds the maximum ZT of 1.1 at 877 K in polycrystalline SnS. Electronic band structure improves SnSe's Seebeck coefficient. The strong anharmonicity of SnSe gives it 0.4 to 0.5 W/m·K thermal conductivity. In p- and n-type crystals, SnSe-based TE excels. New purification methods diminish grain boundary tin oxide. Ultra-pure SnSe has a figure of merit of 3.1 and an average ZT of 2.0 between 400 and 700 K without internal oxygen.

To improve performance by 50%, compare SnSe to ZT (=1) thermoelectrics. Comparison between SiGe nanocomposites with nano inclusions and bulk nanostructured SiGe, $FeSi_2$, and Mg_2Si. Nano–SiGe maximizes power factor and lowers thermal conductivity. $FeSi_2$–SiGe nanocomposites have lower heat conductivity and higher merit. In SiGe–$FeSi_2$, nanoparticles deflect phonons, lowering thermal conductivity. SiGe–Mg_2Si crystal defects and grain boundaries increase phonon scattering centers, reducing thermal conductivity. Sample Mg_2Si's σ rises up to 550°C but decreases with rising temperature. Up to 750°C, SiGe carrier mobility decreases due to electron-phonon scattering. After 750°C, SiGe and $FeSi_2$ samples exhibit decreasing thermal conductivity (k) values due to increased phonon scattering.

Space, band gap, and defect control employing unique synthesis processes are design factors. Scaling and being green require affordable materials. Over 700°C, nanostructured material combinations improve thermoelectric conversion. Designing thermoelectric materials requires improving phonon scattering and electrical conductivity. The material must be cheap, abundant, non-toxic, and scalable [13].

4.3.2 Material Properties for Cooling Applications

Device design and high ZT values enhance thermoelectric material potential for TEC performance. Most TECs weld highly conductive metal electrodes to connect p–n thermoelectric materials in series and thermally in parallel. Au deposits on the thermoelectric legs allow strong electrical and thermal contact between the Au electrodes and legs and a fast response time.

Battery performance may be affected by overheating and working temperature. Batteries perform best at ~298 K, with a recommended working temperature below 323 K. Active battery cooling with new TECs maintains battery temperature at ~303 K, unlike liquid and natural convection [14].

Monitored the temperature differential of module 1 and module 2 while supplying current to evaluate the ZT's improved cooling. Figure 4.6a and b show the outcomes. The Peltier effect heats and cools two semiconductors under DC current. Two modules monitored thermoelectric cooling. Commercial bismuth telluride was utilized for p-type legs in Modules 1 and 2, and $Bi_2Te_{2.69}$–$Se_{0.33}Cl_{0.03}$ and 1 wt% Ag_8SnSe_6 for n-type legs [15].

When a DC current is provided,

1. The Peltier effect decreases the cooling side temperature quickly and constantly until it achieves a constant (Figure 4.6c and d).
2. Module 2 is better cooled than module 1 at same current (2.75 A). Module 2's Tc (2.75 A) is 28% lower than module 1's (283.2 K).

At 2.75 A, Module 2 DT achieved 293.69 K. When cooled, most DTc bismuth telluride couple modules approach 2–5 A saturation. Because thermoelectric elements are temperature-dependent, thermoelectric modules' cooling performance depends on operating temperature (Figure 4.7c and d) [15].

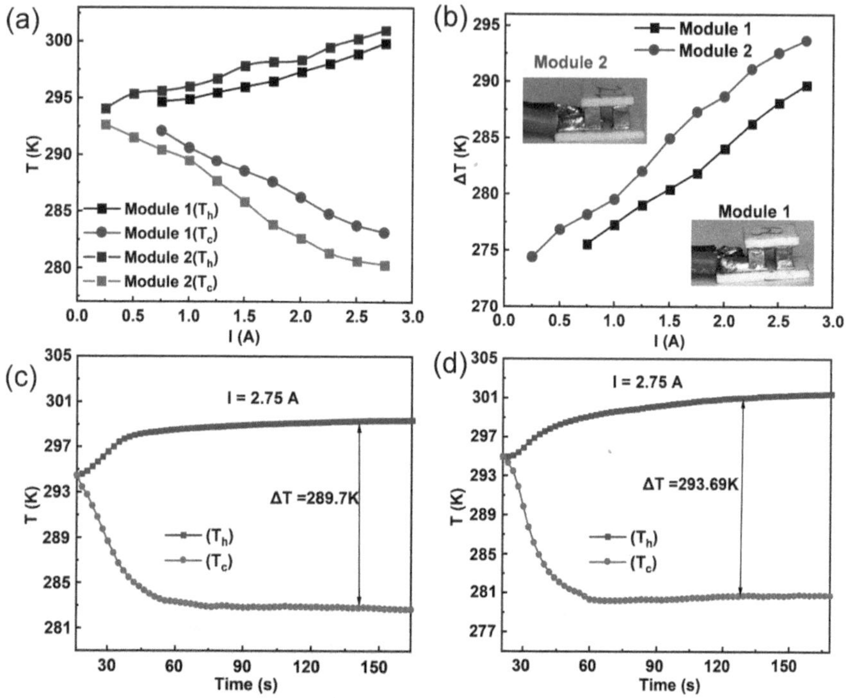

FIGURE 4.6 (a) Electrical current dependence comparison between modules 1 and 2, (b) cooling ΔTc electrical current dependence comparison between modules 1 and 2, (c) module 1 cooling time-dependent effect, and (d) module 2 cooling time-dependent [16].

4.4 PERFORMANCE TRADE-OFFS AND SELECTION GUIDELINES

A static thermoelectric generator has no moving parts. The ends of TE semiconductors generate an electromotive force when their temperatures vary [16].

4.4.1 Body Heat Harvesting

In smart devices and biosensor batteries, thermoelectric materials are popular. Our milliwatt thermoelectric generators are perfect for low-temperature applications, as the human body core is constantly 37°C and the average surface temperature is 33°C. Low-weight, flexible thermoelectric substrates and low heat dissipation limit this use. No solar cells or piezoelectric sensors are needed for thermal generators to supply continuous power. Figure 4.7 shows TEG prototypes targeting garment fibers or wristwatches.

A matched load provides the human-worn TEG with 0.15 V and 0.3 nW open-circuit output. Coating fabric temperature was 300–390 K. Fabric device voltage was 4.3 mV TE at 75.2 K. The studies showed that the upper arm, wrist, chest, and T-shirt had the maximum power.

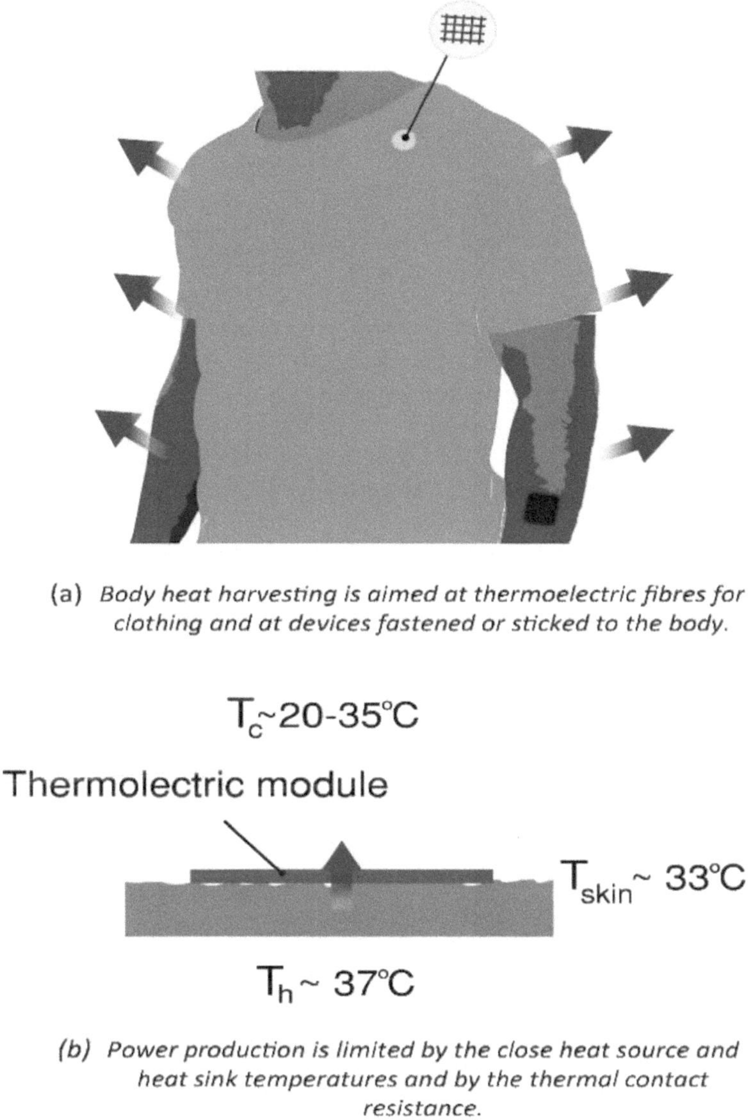

FIGURE 4.7 A thermoelectric generator for body heat gathering. (a) Thermal fibers for clothes and body-attached devices are used for body heat collecting. (b) Thermal contact resistance and near heat source-heat sink temperatures limit power production [17].

A wearable Electrocardiogram (ECG) sensor driven by an upper arm TEG was one option. An upper arm TEG delivered 20 W/cm² of energy at 1.1 m/s. Twelve small thermoelectric modules connected in series and parallel by copper strips form a TEG. The TEG device reached 23 mW and 4.75 mW/cm² during experimental characterization at 35 K. This TEG device generates 10.5 mV of open-circuit voltage when worn on the wrist, powering an LED. TEG fabrication used polymer-based fabric and $Bi_{0.5}Sb_{1.5}Te_3$ and $Bi_2Se_{0.3}Te_{2.7}$ printing. The prototype included 12 thermocouples attached to a 6 by 25 mm^2 rectangle coated in conductive thread. The device outputs 2.7 mV and 8.1 nW at 25°C. Walking states have higher voltages than stationary states because they dissipate heat better. Prototype A was 2.5 mm thick, and prototype B was 1.4 mm.

A multifunctional test thermoelectric generator was used. Generator-temperature plate contact pressure, thermal interface materials, device temperature (same gradient), and flexible generator bend radius were tested. Contact pressure, the most important parameter, affected generation by 10–20 mW, although data suggest a threshold pressure beyond which power output does not increase, mostly due to thermal contact resistance replacing air gaps between surfaces. Thermal interface material alone affected power generation by 51.9% [17].

These devices must also work in a variety of driving scenarios with different exhaust gas temperatures and mass flows. The system's usable exhaust energy, exergy, was analyzed for several driving cycles to determine the optimal system design points. TEG power generation has entered a new phase due to material advances, making this technology more accessible. TEGs will help design engines that convert waste heat into current once materials become cheaper and more abundant. This will reduce gasoline use and CO_2 emissions.

The underside of the car might be subjected to extreme loads such as combustion engine vibrations, road bumps, and winter street salt solutions. To be installed on vehicles in tight locations, TEGs must be small. The most crucial aspect of the TEG system is the cartridges, which must be properly integrated. The cartridges receive exhaust gas, cooling fluid, and electrical connections from the TEG protection. Advances in TEG development, materials, and integration help create vehicle-suitable thermoelectric generators [18].

4.5 CHARACTERIZATION METHODS, STANDARD AND PROTOCOLS

4.5.1 Characterization Methods

Growing technology provides thin sheets of these materials; therefore, their thermoelectric properties must be defined. Large samples simplify thin film in-plane characterization. Under a temperature gradient, open-circuit voltage is measured to determine the in-plane Seebeck coefficient. The measurement system reduces three-dimensional heat flux with accurate temperature and voltage probing. In-plane measurements may favor conductive device layers [19].

Several experiments can measure nano- and micro-film zT. Single transport coefficients are often measured in several configurations using different samples. Few methods can characterize zT on the same sample. The Harman approach characterizes thin film superlattice cross-plane zT values.

Several commercial instruments can measure bulk zT, but thin and thick films are difficult. TE creates an in-plane short circuit on an electrically conductive seed

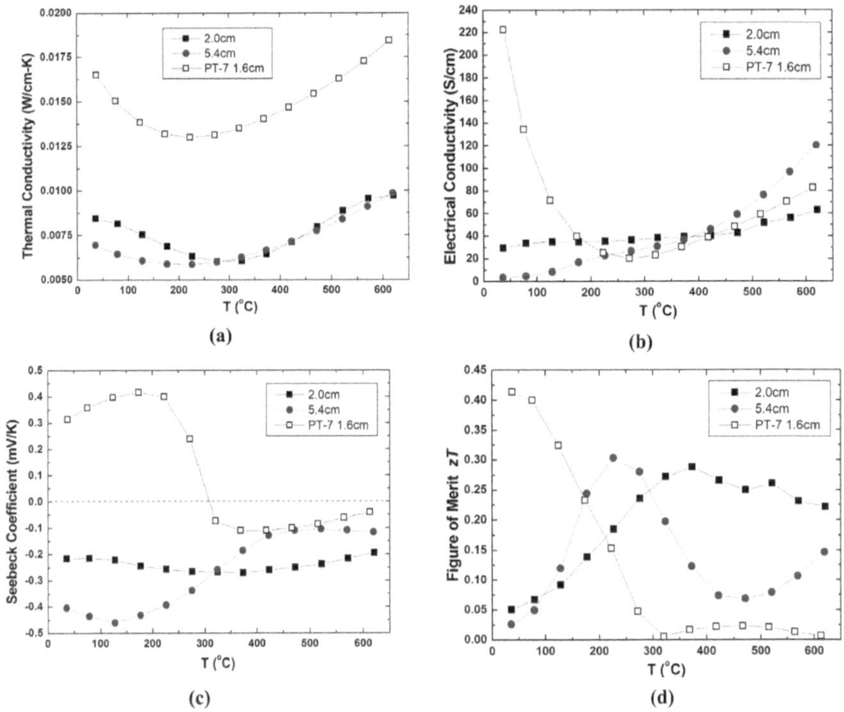

FIGURE 4.8 The measured thermoelectric properties for #1ST: (a) thermal conductivity, (b) electrical conductivity, (c) Seebeck coefficient, and (d) figure of merit, zT [21].

layer in electrochemical films. Eliminating the electrically conductive seed layer allows full in-plane zT assessment of electrochemically generated TE materials [20]. At 130°C, the 5.4 cm sample has a −0.46 mV/K Seebeck coefficient. The figure of merit peaks at 0.30 for various temperature samples (Figure 4.8) [21].

Figure 4.9a shows pure and Al-doped ZnO electrical conductivity from 300 to 1100 K. Every doped sample conducts like metal. Room temperature electrical conductivity is 70,000 S/m for 1% Al-doped material. RF plasma processing enhanced carrier concentration and worked well in Al-doping. Power diminishes as Al dopant concentration grows from 2% to 4%. RF plasma processing enhanced carrier concentration and worked well in Al-doping. Power diminishes as Al dopant concentration grows from 2% to 4%. All samples' Seebeck coefficient temperature dependency is presented in Figure 4.9b. All samples have negative Seebeck coefficients, confirming n-type. Al_3-doped ZnO lattice should add electrons. The Seebeck coefficient dropped by at least one order of magnitude with Al doping.

The Seebeck coefficient decrease may be due to heat pressing. The Seebeck coefficient was lower in hot-pressed samples than in normal sintering. Carrier concentration problems can result from reduced environment sintering. Conductivity rises, and the Seebeck coefficient falls. Figure 4.9c illustrates the sample power

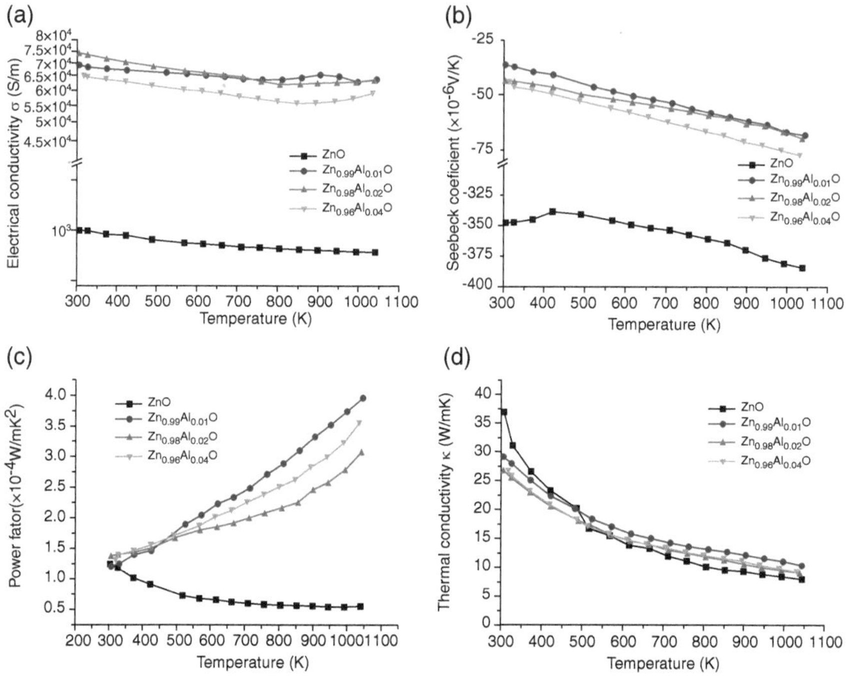

FIGURE 4.9 Temperature dependence of the (a) electrical conductivity, (b) Seebeck coefficient, (c) power factor, and (d) thermal conductivity of the hot-pressed ZnO samples [22].

factor and temperature dependency. Al doping at high temperatures boosts the power factor. Figure 4.9d shows how temperature affects sample thermal conductivity (k). Since grain boundary scattering does not influence heat conductivity, thermal conductivity decreases with temperature in both doped and undoped ZnO. Denser dopants lower ZnO lattice thermal conductivity [22].

4.5.2 Selecting and Characterizing Thermoelectric Materials for Heating and Cooling Applications

Semiconductors, ceramics, and polymers make up most TEs. Recent research found that ethylenedioxythiophene and carbon fiber polymer–matrix structural composites are thermoelectric. Inorganic thermoelectric materials' main problems are toxicity, resource scarcity, and difficult manufacturing. Creating new materials requires improvement. Compared to other thermoelectric modules, insulating polymer matrices and conductive fillers are greener. Further analysis included mechanical flexibility, low-cost manufacturing, solution processing capability, and lightweight design. A discrete conductive polymer composite of carbon nanotubes (CNT) and bismuth telluride was tested for thermoelectric performance, as shown in Figure 4.10.

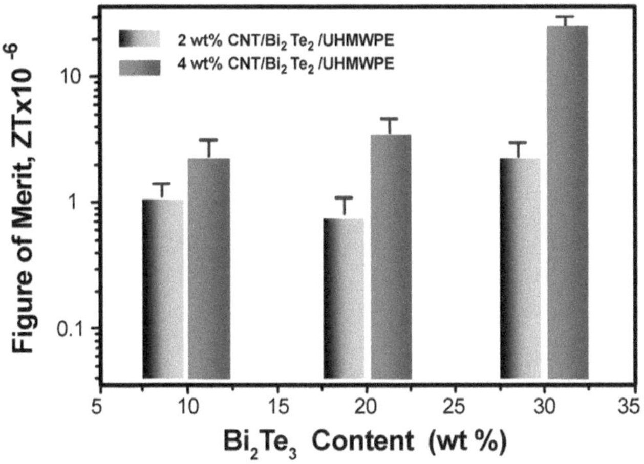

FIGURE 4.10 CNT/Bi_2Te_3/UHMWPE composite ZT with varied CNT and Bi_2Te_3 compositions [23].

A temperature difference between the hot and cold sides develops TEG power. Increasing TH lowers TEG performance, especially at its durable temperature. Power generation decreases with temperature. The optimal temperature maximizes power output. Thermoelectric modules are tested at different flow patterns, heating temperatures, water flow rates, and series numbers. The heat source or heating temperature influences performance more than the heat sink pattern or water velocity. Thus, lower water flow rates reduce power consumption, and higher hot-side temperatures enhance the temperature difference, increasing TEM performance.

4.5.3 Standard and Protocols

Thermoelectric materials have improved to enhance thermal engine and heat-consuming technology energy efficiency, increasing interest in power generation. It's less common to test thermoelectric modules (TEMs) for power-generation performance than for the thermoelectric qualities of new materials. To improve designs and validate implementations, TEM properties must be studied at varied compressive pressures to reduce heat barriers. TEM installation in power-generation systems requires complicated module performance monitoring [24].

Device design, data gathering procedures, and contact geometry affect materials data, making interlaboratory comparison difficult. The Seebeck coefficient must be measured as a function of contact geometry under both steady-state and transient thermal settings to determine how these parameters affect it at high temperatures and to develop standard testing techniques. It is difficult to measure the temperature of the surface that is in direct contact with the object because thermal offsets are common. The thermally distinct contact surfaces of each system with the external environment, probe, and sample result in parasitic heat transfer with the surrounding environment, which in turn causes the surface temperature to be reported as different [25].

4.5.4 Protocol

Rigid analytical data acceptance protocols: The thermoelectric community is advised to adhere to a specific set of guidelines that must be followed when publishing any thermoelectric research at the ITC plenary meetings. Each sample should feed reference material data into the initial property estimation model after testing committee approval. Sample property estimations should improve. In an ideal world, this suggestion would be superfluous. This proposal addresses current thermoelectric research and engineering linkages. This method must develop reference materials for general usage so that scholars and enterprises can agree on a single, authorized collection [26].

REFERENCES

1. Solanki, P. M., & Deshmukh, D. D. (2014). Materials selection criteria for thermoelectric power generators performance enhancement: a critical review. *PRATIBHA: International Journal of Science, Spirituality, Business and Technology (IJSSBT)*, *3*(1), 27–32.
2. Rad, M. K., Rezania, A., Omid, M., Rajabipour, A., & Rosendahl, L. (2019). Study on material properties effect for maximization of thermoelectric power generation. *Renewable Energy*, *138*, 236–242.
3. Tritt, T. M., & Browning, V. M. (2001). Overview of measurement and characterization techniques for thermoelectric materials. In Tritt, T. M. (ed.). *Semiconductors and Semimetals* (Vol. 69, pp. 25–49). Elsevier.
4. Russ, B., Glaudell, A., Urban, J. J., Chabinyc, M. L., & Segalman, R. A. (2016). Organic thermoelectric materials for energy harvesting and temperature control. *Nature Reviews Materials*, *1*(10), 1–14.
5. Bell, L. E. (2008). Cooling, heating, generating power, and recovering waste heat with thermoelectric systems. *Science*, *321*(5895), 1457–1461.
6. Twaha, S., Zhu, J., Yan, Y., & Li, B. (2016). A comprehensive review of thermoelectric technology: materials, applications, modelling and performance improvement. *Renewable and Sustainable Energy Reviews*, *65*, 698–726.
7. Enescu, D., & Virjoghe, E. O. (2014). A review on thermoelectric cooling parameters and performance. *Renewable and Sustainable Energy Reviews*, *38*, 903–916.
8. Mao, J., Chen, G., & Ren, Z. (2021). Thermoelectric cooling materials. *Nature Materials*, *20*(4), 454–461.
9. Ding, J., Zhao, W., Jin, W., Di, C. A., & Zhu, D. (2021). Advanced thermoelectric materials for flexible cooling application. *Advanced Functional Materials*, *31*(20), 2010695.
10. Chein, R., & Huang, G. (2004). Thermoelectric cooler application in electronic cooling. *Applied Thermal Engineering*, *24*(14–15), 2207–2217.
11. LeBlanc, S. (2014). Thermoelectric generators: linking material properties and systems engineering for waste heat recovery applications. *Sustainable Materials and Technologies*, *1*, 26–35.

12. Portions of this document appear in: Zhu, Qing, H. S. K., & Zhifeng, R. (2017). A rapid method to extract Seebeck coefficient under a large temperature difference. *Review of Scientific Instruments*, *88*(9), 094902. And in: Zhu, Q., Shaowei, S., Hangtian, Z., & Zhifeng, R. (2019). Realizing high conversion efficiency of Mg_3Sb_2-based thermoelectric materials. *Journal of Power Sources*, *414*, 393–400.
13. Sudharani, P. V. (2020). A brief review on thermoelectric materials and their applications. *International Journal of Science and Research (IJSR)*. https://doi.org/10.10.21275/SR22119183928
14. Chavan, S., Rudrapati, R., & Manickam, S. (2022). A comprehensive review on current advances of thermal energy storage and its applications. *Alexandria Engineering Journal*, *61*(7), 5455–5463.
15. Haruna, A. Y., Luo, Y., Li, W., Ma, Z., Yang, L., Zhang, Z., ... & Yang, J. (2023). High thermoelectric performance in multiscale Ag_8SnSe_6 included n-type bismuth telluride for cooling application. *Materials Today Energy*, *35*, 101332.
16. Hadjistassou, C., Kyriakides, E., & Georgiou, J. (2013). Designing high efficiency segmented thermoelectric generators. *Energy Conversion and Management*, *66*, 165–172.
17. Fernández-Yáñez, P., Romero, V., Armas, O., & Cerretti, G. (2021). Thermal management of thermoelectric generators for waste energy recovery. *Applied Thermal Engineering*, *196*, 117291.
18. Ranalli, M., Adldinger, M., Kossakovski, D., & Womann, M. (2013). Thermoelectric generators from aerospace to automotive. *ATZ Worldwide*, *115*(9), 60–65.
19. Bahk, J. H., Favaloro, T., & Shakouri, A. (2013). Thin film thermoelectric characterization techniques. *Annual Review of Heat Transfer*, *16*, 51–99.
20. Barati, V., Garcia Fernandez, J., Geishendorf, K., Schnatmann, L. U., Lammel, M., Kunzmann, A., ... & Reith, H. (2020). Thermoelectric characterization platform for electrochemically deposited materials. *Advanced Electronic Materials*, *6*(4), 1901288.
21. Su, C. H. (2019). Design, growth and characterization of PbTe-based thermoelectric materials. *Progress in Crystal Growth and Characterization of Materials*, *65*(2), 47–94.
22. Cheng, H., Xu, X. J., Hng, H. H., & Ma, J. J. C. I. (2009). Characterization of Al-doped ZnO thermoelectric materials prepared by RF plasma powder processing and hot press sintering. *Ceramics International*, *35*(8), 3067–3072.
23. Pourkiaei, S. M., Ahmadi, M. H., Sadeghzadeh, M., Moosavi, S., Pourfayaz, F., Chen, L., & Kumar, R. (2019). Thermoelectric cooler and thermoelectric generator devices: a review of present and potential applications, modeling and materials. *Energy*, *186*, 115849.
24. Hejtmánek, J., Knížek, K., Švejda, V., Horna, P., & Sikora, M. (2014). Test system for thermoelectric modules and materials. *Journal of Electronic Materials*, *43*, 3726–3732.
25. Martin, J. (2013). Protocols for the high temperature measurement of the Seebeck coefficient in thermoelectric materials. *Measurement Science and Technology*, *24*(8), 085601.
26. Lyon Jr, H. B. (1994, August). Validation of thermoelectric measurements. *AIP Conference Proceedings*, *316*(1), 139–141.

5 3D Printing of Thermoelectric Materials and Devices

5.1 INTRODUCTION TO 3D PRINTING/ADDITIVE MANUFACTURING

Materials that convert heat to electricity are thermoelectric (TE) devices. With additive manufacturing, complex structures may be made from 3D model data. Additive fabrication for TE materials and devices is expanding rapidly, despite being less mainstream.

A multistep thermoelectric fused filament fabrication experiment is shown in Figure 5.1. To start sample production, a Filastruder was used to mix and extrude Bi_2Te_3 and ABS powder at 220 °C. Ultimaker 2 printing was done using custom filament. Layer height, infill ratio, and printing speed were 100%, 0.2 mm, and 40 mm/s, respectively. The mass percentage of Bi_2Te_3 to ABS was consistently maintained at 80% in all combinations, as greater ratios resulted in the formation of brittle filaments that broke in the 3D printer tube. The temperature of the 0.4 mm nozzle was maintained at 260 °C. The heat bed was heated to a temperature of 90 °C in order to facilitate material deposition. The initial layer was fastened with double-sided tape. Cubes with 10 mm edges were printed.

Inks, pastes, and other substances are consistently applied onto inflexible or flexible surfaces using either roll-to-roll (R2R) or sheet-to-sheet (S2S) 2D methods. This process deposits materials between the unwinder and winder wheels. To repeatedly make components, R2R substrate-based manufacturing technologies combine additive and subtractive techniques, like laser scribing. R2R uses several technologies and mass production to create completed material rolls efficiently and cheaply.

3D printing creates parts and components from several materials with different mechanical and physical properties through sequential material layer addition [2].

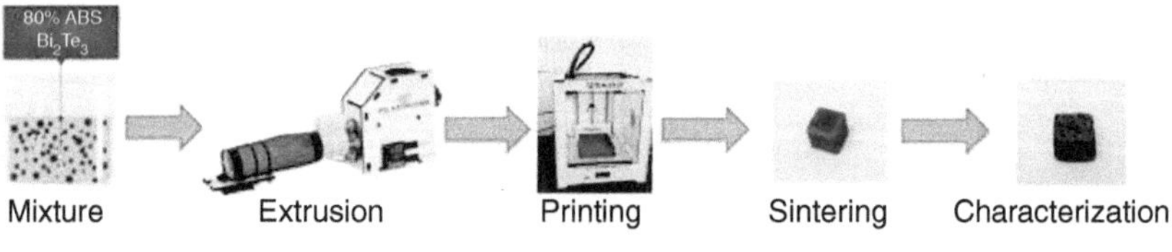

FIGURE 5.1 Schematic of sintered sample production [1].

DOI: 10.1201/9781003639343-5

Bulk ceramics and polymer composites connect piezoelectric constants. Dropping affects the piezoelectric constant but not flexibility or sensitivity. Due to their brittleness, intricacy, and processing difficulty, piezoelectric materials are challenging to handle. Recently, piezoelectric material additive manufacturing researchers have made advancements. The new piezoelectric ceramic material is more elastic due to its design and manufacture [3].

TEG works in many environments because glass can sustain higher temperatures than most flexible substrates. Another unique method is printing on window glass to capture heat. A syringe barrel, automatic pressure regulator, and three-axis precision CNC router make up the lab-developed direct ink write (DIW) system. The substrate and syringe barrel are supported by a CNC router that controls the TEG's 10 mm resolution. Air pressure is regulated. Auto-regulatory syringe barrels hold the chemical. The syringe barrel's metallic nozzle presses material onto the substrate.

A 750 mm metallic nozzle prints TEG at 30 and 40 mm. Experimental parameters for extruding material are matched consistently with pressure, standoff distance, and substrate speed. Printing speed is determined by substrate speed, pressure extrudes material, and standoff distance measures the distance between the nozzle tip and substrate. Figure 5.2a illustrates printing, and samples are dehydrated in a dry oven for 30 minutes at 140°C following printing in order to remove solvent residue. Use of EN-06B8 conductive silver paste defines TEG connections (Figure 5.2b).

Figure 5.1a depicts printing. According to the manufacturer's recommendations, samples are subjected to a drying process in a dry oven at a temperature of 140°C for a duration of 30 minutes in order to eliminate any remaining solvent residue after the printing process. For characterization, TEG connections use conductive silver paste. A computer-interfaced K-type thermocouple monitors both end temperatures while the PID controller adjusts the heat source temperature. GPS-8055C computer-interfaced Digital Multimeters capture TEG output changes.

DIW and FDM layer things by extruding and depositing semi-liquid or molten materials with nozzles or extrusion devices. Extrusion can handle thermoplastics, thermosetting polymers, pastes, bioinks, and metals. A computer-interfaced

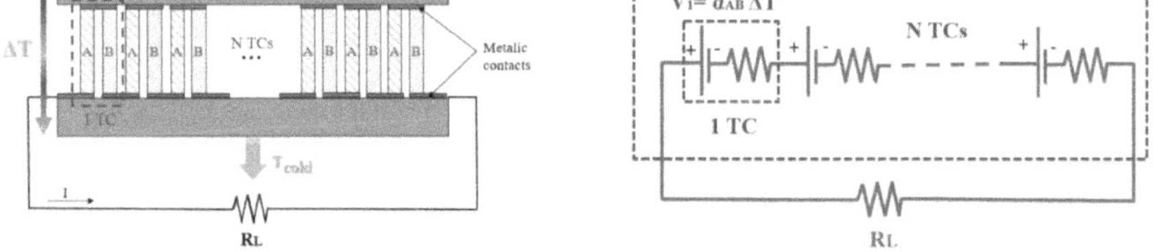

FIGURE 5.2 (a) Schematic of the printing process of TEG and (b) schematic of the characterization setup [4].

K-type thermocouple monitors both end temperatures while the PID controller adjusts the heat source temperature. GPS-8055C computer-interfaced Digital Multimeters capture TEG output changes. For repeatability, every sample is examined five times, and the difference is computed.

Extrusion-based 3D printing has several advantages; however, architecture complexity, resolution, and precision may be limited. Printing polymerizes and solidifies liquid resin by selectively exposing it to UV rays. 3D printing based on SLA and DLP photopolymerization can create precise and intricate objects, but its material system, printing volume, and equipment costs are restricted. Powder bed fusion, also known as powder-based 3D printing, processes powdered materials into 3D objects. Only thermoplastic polymer-bound graphene/polymer composites work.

Understanding 3D printed graphene-based materials, structures, and functionalities is ongoing. Multipurpose materials and complex architectures can be effortlessly combined in 3D printing to build all-in-one bespoke gadgets with new possibilities. Adding microscopic and macroscopic functional features to 3D printed composites can suit several application criteria. Multifunctional sensors for temperature, pressure, bending, tensile strain, and moisture monitoring could be coupled with sensible material and structural design in 3D printing. Integrating and decoupling multifunctional graphene-based materials allow sophisticated applications [5].

The TE legs of these earlier devices featured complicated cross-section patterns but no spatial organization, despite the possibility of 3D printing producing complex structures. Complex bottom designs are developed by stretching them along the perpendicular axis.

TE leg geometry can be optimized via 3D printing to improve RTG and TE generator output in various cases. DIW 3D printing relies on TE slurry printability and material TE. In order for the TE slurry to be printable, it needs to have sufficient rheology to both hold its shape on the substrate and extrude from the needle without clogging. The most crucial aspect to enhance in slurry-derived TE materials is electrical conductivity due to the loose interaction between TE particles and organic binders [6].

A rheologically tailored slurry can be 3D-sculpted. Binder and volatile solvent slurry powder. Polyvinylpyrrolidone (PVP) is a viscoelastic material that binds slurries because of its low melting point and non-toxicity. The slurry must be viscoelastic and shear-thinning. Higher-solids slurry homogeneity and stability affect printing.

3D printing with direct writing, because of TE slurries and printing techniques, has rarely been investigated for TE materials and devices. In 3D printed Bi_2Te_3/PVP and $Bi_{0.5}Sb_{1.5}Te_3$/PVP composites that are directly written and have different TE fillers and heat treatment times, TE is examined.

Output is measured on 3D-printed composite TE devices. This finite element approach study analyzes how TE leg size impacts device output. The materials TE/PVP Composite Slurries: Fillers were 2–10 mm P-type $Bi_{0.5}Sb_{1.5}Te_3$ and n-type Bi_2Te_3 powders. Organic PVP binds with DMF. The TE slurry was made using TE powder, organic binder, volatile solvent, and 60%–91% TE filler. A homogeneous colloidal slurry was made by dissolving and agitating PVP in DMF.

A stable TE slurry was created by adding TE powder to the solution and whirling it at high speed for nearly 10h to evaporate most of the Design for Manufacturing (DMF) and evenly mix it into the colloidal solution. Adventure 3D-LB Printers printed TE/PVP composites. The computer control system regulated the printer needle route and speed. Shear elastic and viscoelastic moduli were obtained by a 1 Hz oscillatory logarithmic stress sweep. Thermal gravimetry and differential scanning calorimetry determined the breakdown temperature of dried TE composites with 91 wt% TE filler. A flow rate of 20 mL/min of argon was employed to monitor a heating rate of 5°C/min from the initial ambient temperature to a final temperature of 550°C [7].

By using several fabrication processes, FTEG research continues. Dispenser, evaporation, lift-off, lithography, etching, screen printing, and sputtering are fabrication technologies. A thin-film evaporation-made FTEG prototype with seven pairs of n- and p-type TE legs at 6.6 K and 2 nW power output was created. With 100 TE connections and RF co-sputtering, their FTEG prototype produced 4 nW at 15 K. Screen-printed polyethylene naphthalate (PEN) substrates with polystyrene created an FTEG with 5 and 108 n- and p-type carbon nanotubes. The prototype has 55 mW/m^2 power density.

Affordable, easy thin-film-free dispenser printing. An FTEG with 8 TE pairs of n- and p-type Bi_2Te_3 on PDMS was dispenser-printed. The 19 K prototype produced 2.1 mW. One epoxy composite thin-film FTEG with 60 TE elements produced 20.5 mW at 20 K using identical technology. This prototype has 55 mW/m^2 power density.

The polymer fabric was utilized to create FTEG prototypes with 12 and 20 pairs of n- and p-type Bi_2Te_3. FTEG prototypes at 27 and 15 K produced 178 and 224 nW. Dispenser printing requires high heat resistance, thin, flexible substrates, regulated dispensing equipment, and high curing temperature, making it prohibitively expensive and sophisticated for broad application. Manually prepare the substrate, sew the n-leg and p-leg, and fill the TE paste, except for curing. This structure makes a wearable and flexible FTEG for body thermal energy collection [8].

5.2 THERMOELECTRIC MATERIAL FORMULATIONS FOR 3D PRINTING

Hall effect measurements on 3D-printed $Bi_xSb_{2-x}Te_3$ (x=0.3–0.6) samples confirmed composition engineering carrier concentration control. The 3D-printed samples' hole concentration (nh) decreases monotonically with increasing x, from $4.56\times10^{19}cm^{-3}$ for $Bi_{0.3}Sb_{1.7}Te_3$ to $1.28\times10^{19}cm^{-3}$ for $Bi_{0.6}Sb_{1.4}Te_3$. As x increased from 0.3 to 0.5, hole mobility (μh) grew from 173.02 to 276.511 $cm^2/V{\cdot}s$, but decreased to 178.02 at 0.6 (Figure 5.3a).

Semiconductor doping increases carrier–carrier dispersion and decreases carrier mobility of 3D-printed $Bi_xSb_{2-x}Te_3$ (x=0.3–0.5). The hypothesis posits that the current behavior of μ_h is a result of a conflict between scattering and

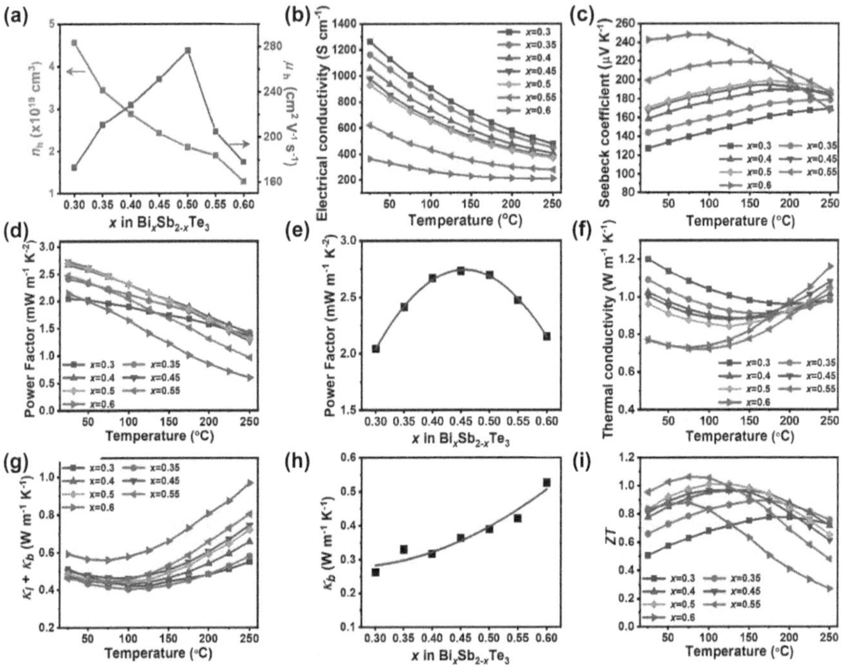

FIGURE 5.3 $Bi_xSb_{2-x}Te_3$ 3D-printed TE characteristics: (a) 3D-printed material room temperature hole concentrations and mobilities, (b) electrical conductivities, (c) Seebeck coefficients, (d) power factors of 3D-printed materials are temperature-dependent, (e) composition-dependent room-temperature 3D-printed material power factors, temperature-dependent (f) and (g) lattice (κl) and bipolar (κb) thermal conductivities, (h) bipolar thermal conductivity of 3D-printed materials at 250°C based on composition, and (i) 3D-printed material ZT values vary with temperature [9].

charge transfer (Figure 5.3b). Electrical conductivities declined with temperature in all samples, characteristic of degenerate semiconductors as lattice vibrations scatter electrons (Figure 5.3c). The reciprocal link between carrier concentration, electrical conductivity, and Seebeck coefficient explains this composition-dependent electrical behavior. Higher μh reduced the bipolar contribution, raising the maximum Seebeck coefficient temperatures with decreasing x.

Sample power factors peak at 27.3 $\mu W/cm^2{\cdot}K^{-2}$ at x=0.45, indicating optimal nh for this 3D-printed material (Figure 5.3d and e). The 3D-printed materials' thermal conductivity was composition-dependent (Figure 5.3f). Thermal conductivity in $Bi_xSb_{2-x}Te_3$ decreased from 1.2 to 0.77 W/m·K at ambient temperature as x increased. $Bi_{0.55}Sb_{1.45}Te_3$ has the lowest thermal conductivity (0.72 W/m·K) at 348 K, compared to bulk crystalline BiSbTe (1.5–2.5 W). Higher x enhances composition-dependent, temperature-dependent thermal conductivity (Figure 5.3g).

Nanostructured TE BiSbTe materials may exhibit low lattice thermal conductivities due to many multiscale pores scattering phonons over a large wavelength range (Figure 5.3g). Composition-engineered, 3D-printed $Bi_xSb_{2-x}Te_3$ materials

(Figure 5.3h) showed peak ZT changes from ambient to 250 °C with increasing x (Figure 5.3i). At 75 °C, $Bi_{0.55}Sb_{1.45}Te_3$ had the greatest ZT of 1.06. At 25°C, $Bi_{0.35}Sb_{1.65}Te_3$ had a peak ZT of 0.76, three times greater than $Bi_{0.6}Sb_{1.4}Te_3$ at ambient temperature. 3D-printed composition-engineered TE samples exhibit greater maxima than bulk-BiSbTe ingots (ZT=0.8–1.0) and approximate nano-structured materials (ZT ~ 1.1–1.9).

High aspect ratio devices will be more practical in real conditions than high power factor and thermal conductivity materials. 3D printing simplifies thermoelectric device construction, which requires precise electrical connections, leg metalization, and thermal and electrical insulation. The lack of high zT values, along with uncomplicated manufacturing techniques, could impede the commercialization of thermoelectric technology [10].

A modified squeegee holder and guideways attachment reduced screen printing pressure and angle (Figure 5.4). Tested ink film thickness, thermal conductivity, Hall effect, and TEG.

CNTs are cylindrical carbon allotropes with excellent mechanical and electrical properties. In bulk (>10 layers), graphene has a mobility of 10,000 $cm^2/V{\cdot}s$, and >140,000 in monolayer form. The best polymer matrix fillers are CNTs and graphene. SWNTs are prospective CNTs due to their high Seebeck coefficient. Despite CNTs' thermal conductivity of ~600 W/m·K, their large surface area can enhance phonon scattering and decrease thermal conductivity by producing more interfaces (polymer/nanoparticle or crystalline/amorphous). Additionally,, thermoelectrics are polymer/spherical PbTe nanoparticles.

Screen, inkjet, dispenser, spray, and spin coating using solution-processed polymer composite materials are cost-effective approaches to mass-produce TE devices. The most difficult aspect of polymer/CNT composite manufacturing is CNT dispersion control. High surface energy and van der Waals attractive forces cause nanotubes to clump in polymers, decreasing electrical conductivity.

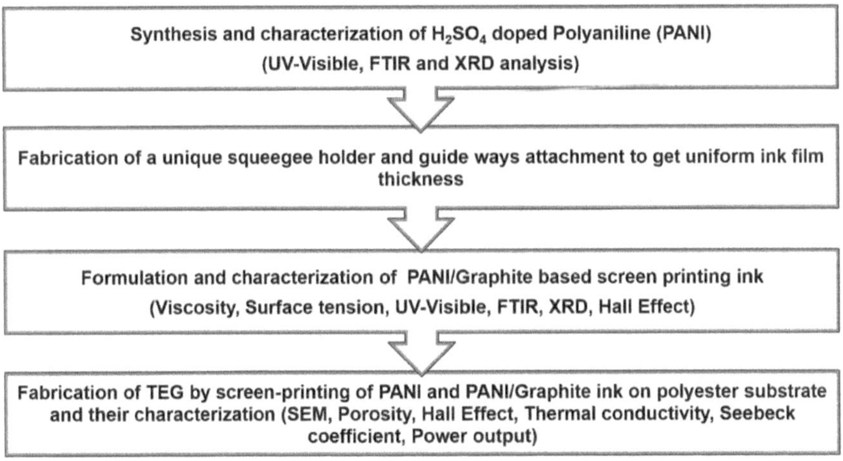

FIGURE 5.4 Methodology of preparing TEG device [11].

PEDOT: PSS increased composite electrical conductivity by allowing charge carriers to travel. Electrical conductivity rose to 40,000 1/Ω·m in the composite with 35 wt% CNT, whereas thermopower and thermal conductivity remained unchanged. CNT loading enhanced electrical conductivity, but thermopower remained constant at 15–30 μV/K. Also, stabilizer type and concentration considerably affected junction carrier transit.

With CNT filler, the polythiophene derivative P3HT's extremely variable molecular structure is intensively researched for thermoelectric properties. It dissolves better in organic solvents than conjugated polymers. TE properties were analyzed on 1–4 μm thick films. Increased electrical conductivity and polymer crystallization on nanotube surfaces result from significant π-stacking between P3HT and CNTs. CNTs were added to P3HT, which had a low σ of ~1.7 ± 0.1 × 10–5 S/cm, causing a rapid increase in electrical conductivity.

Electrical percolation results from carbon nanotubes' low weight %. P3HT and PDPPSe polymers affect SSWCNT microstructure and thermoelectricity. PPPSe and P3HT, with unique molecular structures and electrical characteristics, were used to create nanocomposite films (~25–35 μm thick). Compared to the P3HT/SSWCNT nanocomposite, the PDPPSe/SSWCNT nanocomposite showed better electrical conductivity (537.7 S/cm) and Seebeck coefficient (62.5 μV/K).

Increased PDPPSe/SSWCNT nanocomposite P.F. exhibits 210 μW/m·K² thermal conductivity, 10 times greater than the P3HT/SSWCNT film. PDPPSe's planar form and higher π–π interaction than P3HT enable efficient wrapping around SSWCNT surfaces and compact microstructures. PDPPSe's crystalline orientation aligns bundles in the same plane, improving SSWCNT nanocomposite film electrical conductivity. The energy barrier at the polymer/SSWCNT interface is lower with PDPPSe's 1.32 eV band gap than P3HT's 1.93 eV, allowing charge carrier transfer and improving conductivity (Figure 5.5, Bottom row).

Solution-mixed composites exhibit superior thermoelectric properties. Increased MWCNT concentration led to a Seebeck coefficient of 27.7–22.7 μV/K and a small increase in thermal conductivity. The composite's electrical conductivity increased from 2.34 to 2982 S/m as MWCNT content increased from 10 to 90 wt%. Increasing the quantity of multi-walled carbon nanotubes (MWNT) developed electrically conductive pathways in composites, enhancing electrical conductivity. Temperature enhances the composite's electrical conductivity, implying metallicity. The composite containing 80% by weight of multi-walled carbon nanotubes (MWCNT) exhibited the greatest figure of merit (zT) of 8.71 × 10–4 at a temperature of 120°C.

Conventional thermoelectric device manufacturing begins with ball milling or melting/heating element powders. Alloy powders make ingots, dice, and metallization. Insulating substrate electrical shunts hold TE "legs". The typical production procedure limits TE leg size, shape, and density and may chip or microcrack fragile materials. Thin TE materials with superlattices and nanowires are made using semiconductor fabrication.

Scalability means manufacturing several devices efficiently with minimal material waste and easy design or size changes. Merits compare material and

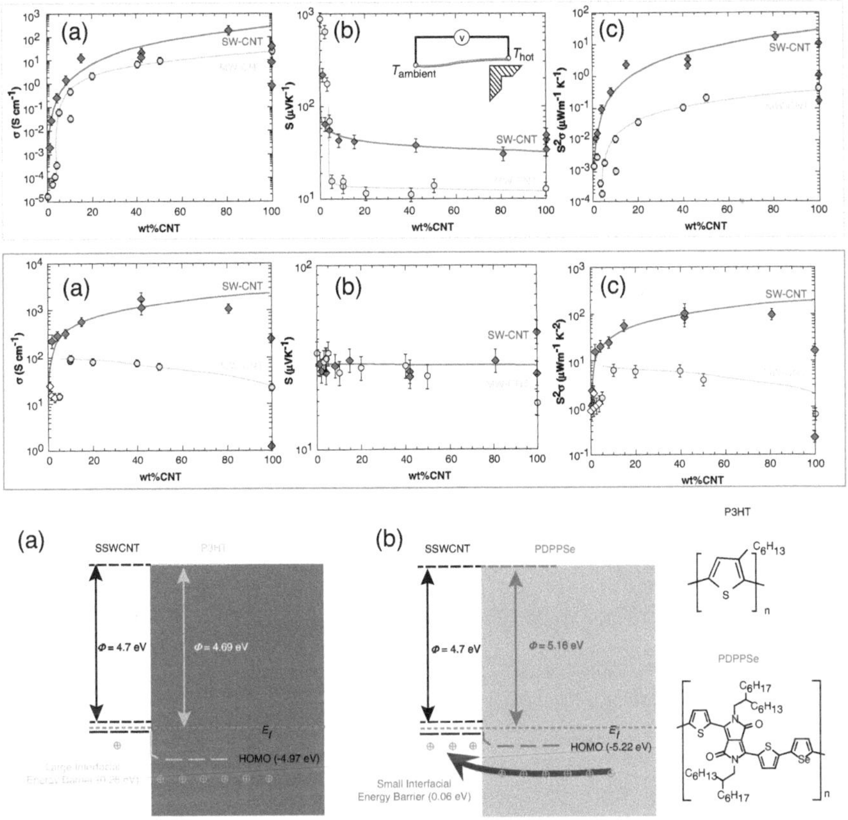

FIGURE 5.5 Top row: TE characteristics of virgin SWCNT (diamonds) and MWCNT (circles) P3HT/CNT mixes. The electrical conductivity, average Seebeck coefficient, and power factor $S^2\sigma$ are shown. TE characteristics of FeCl3-doped P3HT/CNT blends using SWCNT (diamonds) and MWCNT (circles): (a) electrical conductivity, (b) average Seebeck coefficient, and (c) power factor S2σ. The doping time was 1 h with 0.03 M $FeCl_3$ in nitromethane [12].

gadget creation methods. Many inkjet and screen printing studies used power factor, $S^2\sigma$, instead of thermal conductivity. ZT is calculated using the power factor and an approximate thermal conductivity generated from a weighted average of printed material components. Reviews examine inkjet, dispenser, screen, and stereolithography's history and operation [13].

5.3 FDM TECHNOLOGY

Fused Deposition Modeling (FDM) stands out among other 3D printing methods due to its uncomplicated equipment and procedures, cost-effectiveness, wide variety of materials, and high printing precision. FDM utilizes thermoplastic filaments that are printed.

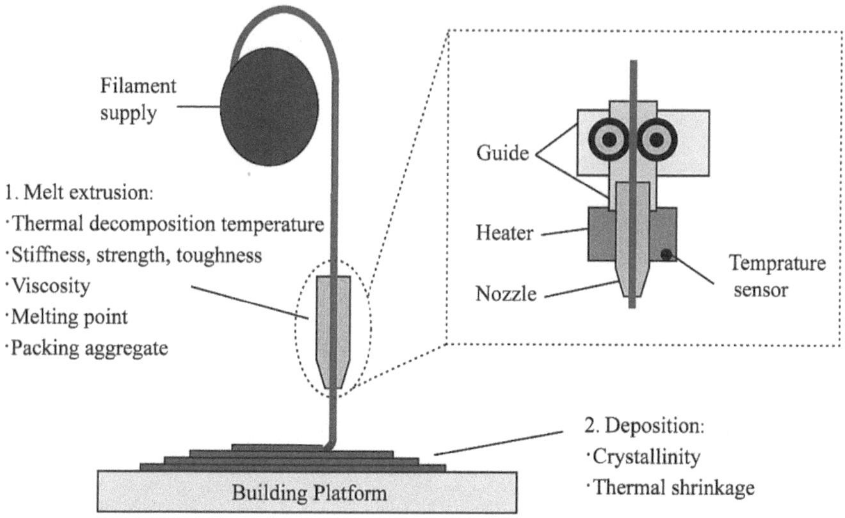

FIGURE 5.6 The process and material requirements of FDM [14].

5.3.1 FDM has Following Features

1. Uninterrupted material extrusion from the nozzle necessitates smooth flow and the absence of large filler clumps.
2. The material must possess high potency, rigidity, hardness, thermal stability, etc., to maintain its shape during molten extrusion.
3. It bridges divides.
4. Low thermal shrinkage and viscosity are required to maintain the shape of the product as it cools.

Figure 5.6 shows FDM material performance criteria. Rheology tests determine printability. Measurements and data sets were needed to characterize polymers' extrusion rheology (Figure 5.7). Layer-by-layer FDM parts make interlayer bonding difficult since their strength impacts part quality. Surface stress causes FDM filament bundles to neck, diffusing and entangling polymer chains (Figure 5.8). Polymer chains of adjacent filaments must diffuse and entangle to form positive interfacial bonding, but filaments solidify quickly, creating interlayer spaces and requiring amorphous and semicrystalline polymers. The FDM process settings considerably impact product quality. Printing quality is hard to control due to the complexity of process parameters.

5.4 STEREOLITHOGRAPHY (SLA)

3D printing allows complex shapes because polymers fuse layer by layer. Additionally, AM methods use a variety of filament, powder, and resin polymers. FDM, SLS, and stereolithography are the main polymeric 3D printing methods. The method of operation, dimensions, surface roughness, resolution, building time, post-processing, usability, and cost are all factors that are compared.

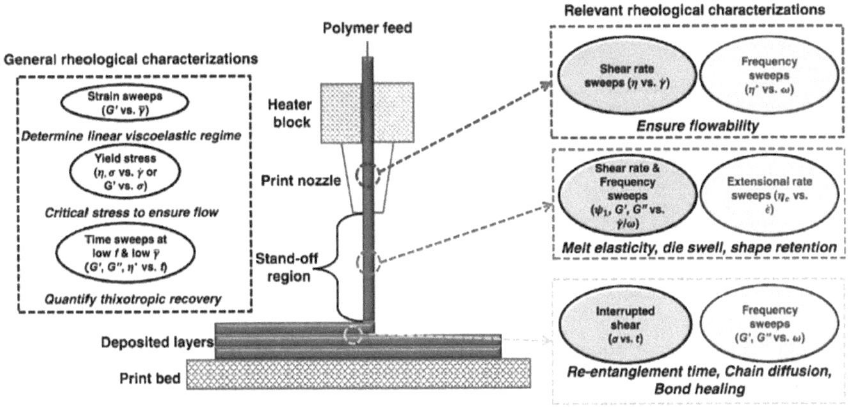

FIGURE 5.7 Rheological measurements and data sets should describe polymer melt behavior at different extrusion phases [14].

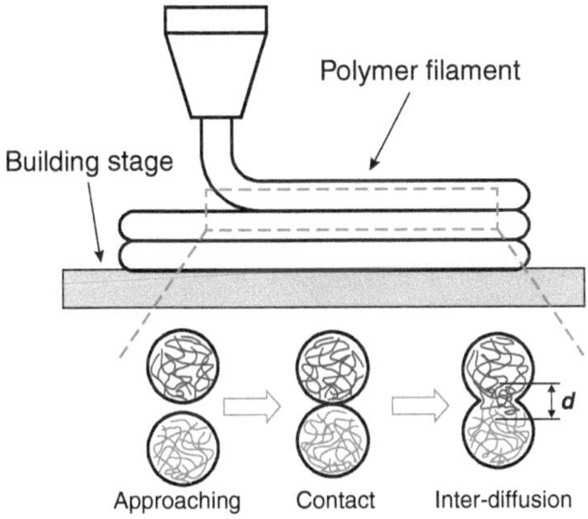

FIGURE 5.8 FDM inter-molecular diffusion between polymer fibers [14].

The printer has a 13.6" LCD screen that can handle 300×298×164 mm prints with a resolution of 46 μm xy and a pixel size of 6480×3600 (7K). In order to adhere to the manufacturer's guidelines for bottom exposure and exposure time, a 50 μm layer thickness was employed. Equipped with a resin tank, build platform, UV lamp, and computer control unit, these 3D printers are quite remarkable, as depicted in Figure 5.9.

5.5 SELECTIVE LASER MELTING (SLM)

Electrons and holes convert heat to electricity as a working fluid in thermoelectric generators. Large-volume thermoelectric module manufacture is inappropriate for hot press and spark plasma sintering due to processing and assembly times. Laser melting (SLM, Figure 5.10), a new thermoelectric material fabrication

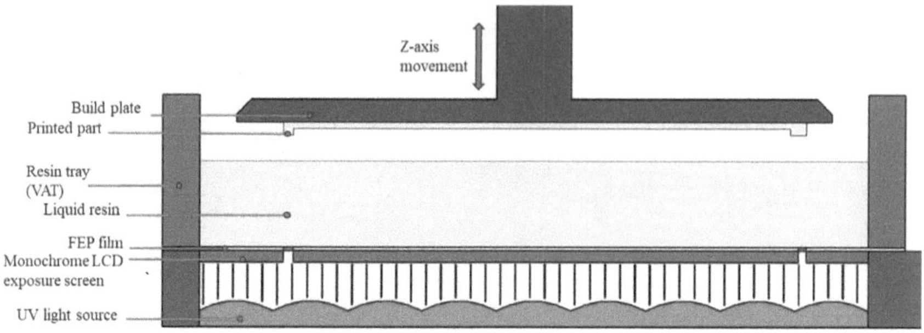

FIGURE 5.9 Description of SLA process [15].

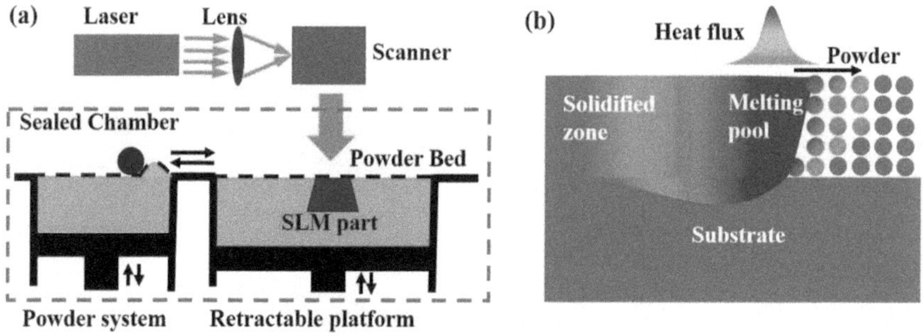

FIGURE 5.10 (a) The working principle of SLM and (b) SLM fabrication of powders [16].

method, can make any shape component. The SLM powder chemical composition can be changed by additive quantities for thermo-element functional doping, which enhances thermoelectric module conversion. A high-power laser fuses particles into a solid with the desired characteristics in SLM (Figure 5.10a).

SLM is ideal for making high melting point, high density materials without adhesives due to its energy density. These properties make it perfect for thermoelectric material production. SLM fabricating thermoelectric materials rapidly converts particles from solid to liquid to solid (Figure 5.10b). Laser movement separates liquid and solid. Even a small amount of oxygen can oxidize SLM in an inert gas atmosphere.

When temperatures reach 1500°C, significant evaporation might damage the physical properties of deposited materials. Because powders are partially solid and liquid, the melting pool edge is distinct from the powder bed. Temperature change and dwell time considerably impact product properties. Power, scan speed, spacing, powder characteristics, chemical composition, laser wavelength, and power affect component density, surface quality, and microstructures.

Compare additive with subtractive or formative manufacturing in Figure 5.11. SLA, FDM, SLM/SLS, and solution printing are additive manufacturing methods, and the devices used are enhancing. The additive manufacturing types and their properties are listed in Table 5.1.

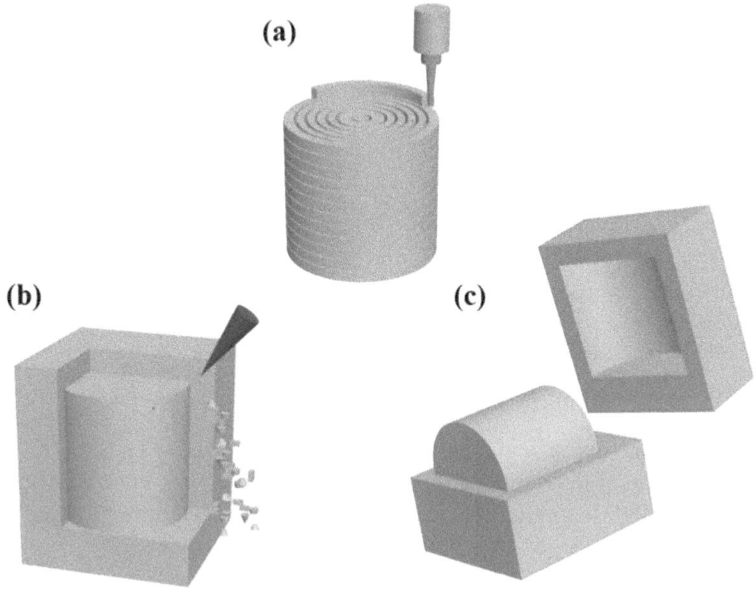

FIGURE 5.11 Comparison of (a) additive, (b) subtractive, and (c) formative manufacturing techniques [17].

TABLE 5.1
Additive Manufacturing Principles Types and Their Properties [17]

Technique	Principle	Consumable Materials	Advantages	Disadvantages
SLA	Use laser scanning to overlay photoresins	Photoresins	A complex internal structure can be produced	Few photoresins exist
FDM	Melt and layer a solid thermoplastic filament to make the desired shape	Solid thermoplastic filaments: acrylonitrile butadiene styrene, polylactic acid, etc.	Support/platform unnecessary	High temperature, nozzle cloggi, filament preparation
SLM/SLS	To sinter particles into the desired shape using selective high-energy lasers	Polymer powders, metal powder	No support needed	Surface quality and speed must be balanced, expensive
SP	Printing the shape layer by layer with inks	Putty/viscous strands in solution	Compatible with polymers, inorganic polymer composites, and inorganic inks	Final goods may shrink

5.6 DESIGN AND OPTIMIZATION OF 3D-PRINTED THERMOELECTRIC DEVICES

According to the 1D Fourier equation of heat conduction, heat moves down the TE legs to generate electricity [18]:

$$\frac{d^2t}{dx^2} + \frac{\varnothing}{\lambda} = 0 \tag{5.1}$$

where

t = Temperature of differential element,
x = Distance that heat flows,
∅ = Heat generated, and
λ = Heat transfer coefficient.

At different heat source temperatures, geometrical TE modules perform better. The consistency of experimental and modeling data implies that increasing the side area and heat dissipation in TE legs can improve output performance. Helix- and spoke-shaped TE legs can be 3D printed, but more complex geometrical shapes cannot [18].

Thermal-element length and area, load resistance, and hot- and cold-side temperatures are used to quantify TEG device efficiency. The nonlinearity of the Thompson heat term limits analytical TEG device efficiency calculations to temperature-independent thermoelectric properties. All FGTM/segmented TEG properties are temperature-dependent, including Seebeck coefficient, electrical resistivity, and thermal resistivity.

By optimizing the TEG's form using sophisticated materials, Fourier's heat conductivity, Peltier effect, Thompson effect, and Joule heat, its highest efficiency was estimated. There are various mathematical segments of the thermo-element. To enhance efficiency at different temperatures, the segment cross-sectional area and length are customized.

The thermo-element's electric current density is [19]

$$J = \frac{I}{A} \tag{5.2}$$

Ohm's law compensates for Seebeck effect's electrical field in thermoelectric materials. Positional electrical field in thermo-element is

$$E = \alpha \nabla T - \rho J \tag{5.3}$$

The cross-section heat transport is calculated as

$$Q = \alpha TJ + k\nabla T \tag{5.4}$$

Energy conservation rule requires the source term from Joule heat to account for heat flux divergence, giving electrical power density as

$$P = \nabla Q \tag{5.5}$$

The largest efficiency is given by [19]

$$H_{\eta Max} = \frac{\sqrt{1+ZT}-1}{\sqrt{1+ZT}+1} \tag{5.6}$$

Ink fluid dynamics and formulation affect print quality, speed, and end-use; hence, printing technologies have improved with ink materials. Ink flow depends on the printing method, equipment, and processing circumstances, making the printing process optimization complex. In all printing methods, no universal ink delivers high-quality prints. Traditional printing causes ink deformations from 10^{-3} to $10^{5}\,s^{-1}$, including distribution, compression, extension, shearing, and transfer.

The challenging flow requires low viscosity inks to avoid significant frictional heat and low productivity, but suitable stiffness after printing. Many researchers prefer to investigate linear and nonlinear viscoelastic interactions before choosing raw materials and printing methods since ink flow parameters are difficult to modify. Recently developed additive manufacturing technologies have expanded industrial and laboratory printing. 3D printing uses organic and inorganic materials such as ceramic-based colloids, polymer composites, graphene oxide hydrogels, and biomaterials to create complex objects with specific geometry.

To investigate 3D printed ink flow behavior without a theoretical framework, simple rules of thumb are used instead of rheology. In order to prevent structural collapse, 3D printable inks have well-structured behavior, high yield stress, and fast elastic recovery kinetics. Three-dimensional printer inks are designed using two methods. A fast-gelling low-viscosity ink is one alternative. Concentrated, solid-like ink that resists deformation is another. Shape retention, green body density, and final product performance are higher with concentrated ink, but its high viscosity and intricate flow behavior demand an exact nozzle diameter, extrusion pressure, and temperature.

Additive manufacturing inorganic colloidal ink composition must consider many factors: customized rheology for low-pressure extrusion through a fine nozzle, rapid solidification after deposition to create the desired structure, and a high colloidal particle volume fraction (0.45–0.50). Concentrated colloidal inks have two additive manufacturing advantages over low viscosity inks. First, a high solute proportion and low polymer percentage in inorganic colloidal inks boost green body density, decreasing drying shrinkage and breakage.

The shear-thinning ability of concentrated colloidal inks permits extrusion through a capillary nozzle and rapid phase shift from a flowable fluid to an elastic solid-like body by eliminating dispersed pressure to retain the pre-designed 3D structure after deposition. Despite their excellent structuring behavior, gravity and colloidal microstructure evolution after deposition can deform and collapse 3D printed structures. The inks' colloidal structural behavior should be studied rheologically before additive manufacturing to understand structural recovery and evolution.

Modeling concentrated colloidal systems as simple shear-thinning fluids with finite yield stress cannot explain die entry and exit during extrusion. Shear rate and time alter the colloidal aggregate microstructure, as well as the elastic and viscous properties, according to a new rheological model.

Due to the rheological importance of ink design, recent 3D printing studies have examined steady and dynamic shear viscosities, storage and loss moduli, and thixotropic behavior to analyze deposited ink gelation, polymer composite and polymer/ceramic blend 3D printability, graphene oxide's binder role, and fused filament fabrication of a solid fiber [20].

A unique piezoelectric platform generates energy from radial artery pulsations. The energy harvester is electrically insulated from the body and shielded from bodily fluids by an air compartment between an elastic sheet and a piezoelectric nanogenerator. The artery expands and compresses the chamber during systole, activating the PENG. The model passed computational and experimental tests. A PENG enclosure and elastic film were readily and cheaply printed using the Ultimaker S5 3D printer with FFM. Several 3D printer settings were evaluated to improve print quality and reduce film thickness [21].

Thermoelectric and other conductive inks or pastes are printed on flexible substrates to make TEGs. Two 2D-printed thermoelectric architectures exist (Figure 5.12). Material layers and heat flow cross-plane in planar construction (Figure 5.12a). Printing materials appear close to each other, and folding them causes in-plane heat flow.

Polymer foil replaces passivated metal foil and ceramic plates in the folded architecture to improve performance. A geometrically constructed device with high zT printed thermoelectric materials performs well. Device efficiency, output power, and effective ZT value depend on the fill factor—the ratio of all TE layer cross-section areas to the device's total cross-section area—and the TE elements' (legs) cross-section areas and lengths. Printed TEGs' fill factor is more important

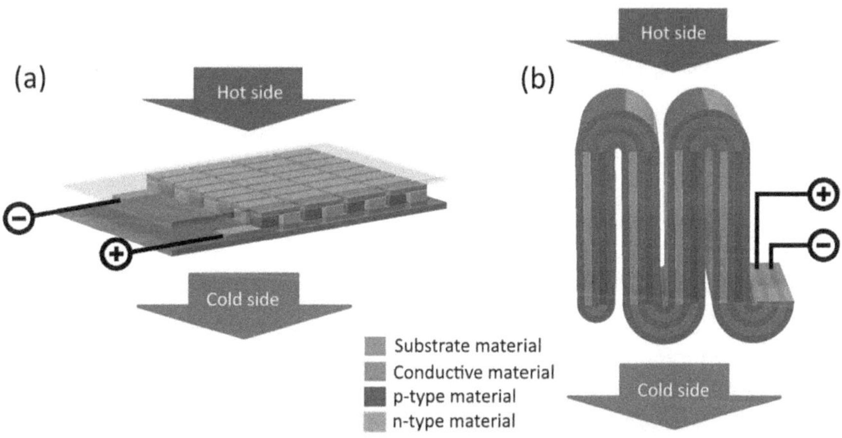

FIGURE 5.12 Model for printed thermoelectric generators. (a) Thin-sheet TEG with substrate in series to thermoelectric material. (b) Thermally parallel substrates and origami folded TEGs with corrugated device architecture [22].

than in traditional devices due to the filler material and substrate heat conductivity. Printer layer thickness, foldable leg length, and application geometry can limit printed TEG thickness. So, printed TEG optimization and design require fill factor freedom.

Analytical optimizations of traditional TEGs sometimes use air as filler, which has a low thermal conductivity of 0.015 $W/m^{-1} K^{-1}$. Small compared to TE materials' thermal conductivities, this quantity is often ignored (rounded to zero). This approximation fails for printed TEGs (Figure 5.12b).

According to Figure 5.13a, a device with 100% fill factor yields the highest output power for both applications due to its optimal Z values. Without fillers, devices must be printed planarly. Fill factor aids processing. For F=0.1, Figure 5.14b shows the same dopt with air (κ_{filler}=0.015 W/(mK)) as the filler material.

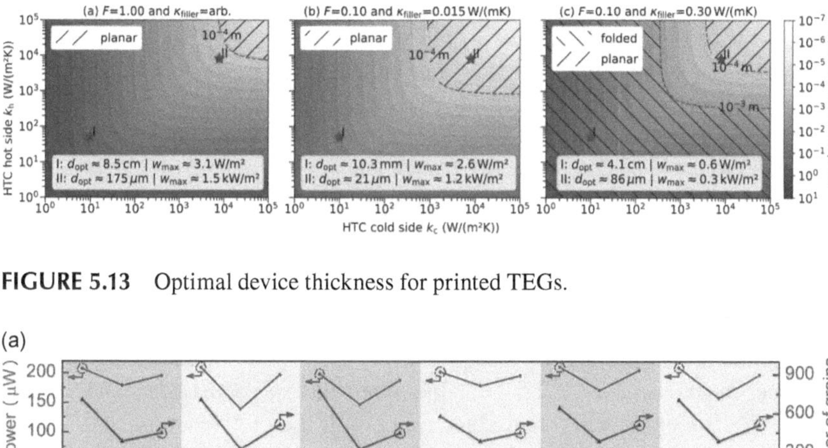

FIGURE 5.13 Optimal device thickness for printed TEGs.

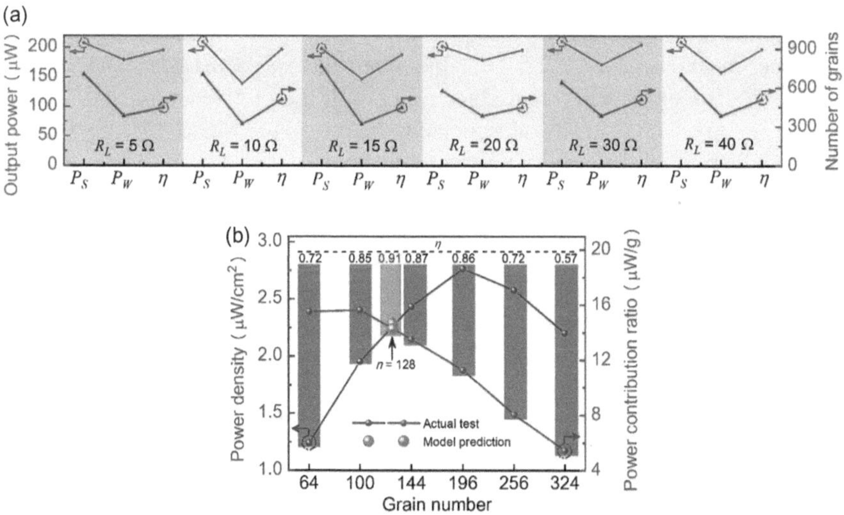

FIGURE 5.14 (a) Comparison of three optimization results utilizing η, PS, or PW criteria under varying loads. S0=4×16 cm^2. The abscissa indicates the optimization objective, the left ordinate represents the f-TEG output power, and the right ordinate represents the number of grains. (b) Comparison of model-based optimization with f-TEG measurements.S0=4×4 cm, R_L=10 Ω [23].

Lowering the fill factor enables a thinner dopt, resulting in a 21 μm thermally impedance-matched planar device for heat exchangers. Low κfiller limits output power loss to 1.2 kW/m². The gadget uses 1.2% less TE material than F=1 to provide this output power.

For F=1, an 8.5 cm thick device performs best but is too big and uncomfortable. At F=0.1, the device thickness is reduced to 10.3 mm, making it more comfortable to wear while still performing effectively at 2.6 W/m². The planar architecture's maximum layer thickness is exceeded by this slimmer device. Low thermal contact resistance applications require folding to increase device thickness. A thermally parallel substrate significantly improves κfiller. In Figure 5.14c, the ideal device thickness is F=0.1 and k_{filler}=0.3 W/(mK).

Heat Transfer Coefficient (HTC) variants for 1 mm-thick foldable smartphones were also shown. Greater k_{filler} and low fill factor dramatically lower output power to 0.6 W/m² and 0.3 kW/m². Increased effective κ results in optimal device thicknesses of 4.1 cm and 86 μm.

Device thickness optimization based on heat transfer coefficients (HTC) for various fill factors and materials. (a) F=1. $k_{filler=}$arb., (b) F=0.1, $k_{filler=}$0.015 W/(mK), and (c) F=0.1, k_{filler}=0.3 W/(mK). Red stars indicate wearable energy-harvesting and waste heat recovery in heat exchangers. The green text boxes show optimal device thicknesses and output power densities. Hatched sections indicate which applications can be accomplished with planar and foldable device architectures [22].

Three optimization outcomes for different load resistances are shown in Figure 5.15a. Although expensive (grain number), PS maximizes TEG power. PW optimization minimizes material cost but provides the lowest TEG power.

Experimental flexible TEG (f-TEG) measurements are compared to optimization model predictions to verify dual-objective optimization's efficacy. f-TEGs feature a 4×4 cm² area and a 10 Ω load resistance (Figure 5.14b). Model-based optimization is effective because the experiment aligns with the theory. The optimization model we use aims to enhance the performance of the wearable monitoring system by optimizing the thermoelectric grains, fill factor, and series-parallel connection mode of the f-TEG. This is done to meet the load requirements more effectively.

Topology optimization optimizes performance characteristics, such as sink base plate average temperature, to distribute material repeatedly in a design domain. An optimized voxel-based void-solid distribution of zeros and ones is produced. Gradient-based optimization works well for binary optimization problems with continuous density distributions. The final steps involve numerical simulations, regularization, and mathematical programming design improvements.

An advection-diffusion heat transfer equation optimizes the stable laminar incompressible fluid model. In horizontal and vertical optimization, the Finite Element Method (FEM) discretizes partial differential equations on a regular grid of 384×640×0 (half domain) and 192×64×0 (quarter domain). Modeling accurately and allowing tiny aspects in the finest design requires proper resolution. Each element's constant density field is used to iteratively alter discrete values in gradient-based optimization, a parallel method of Moving Asymptotes.

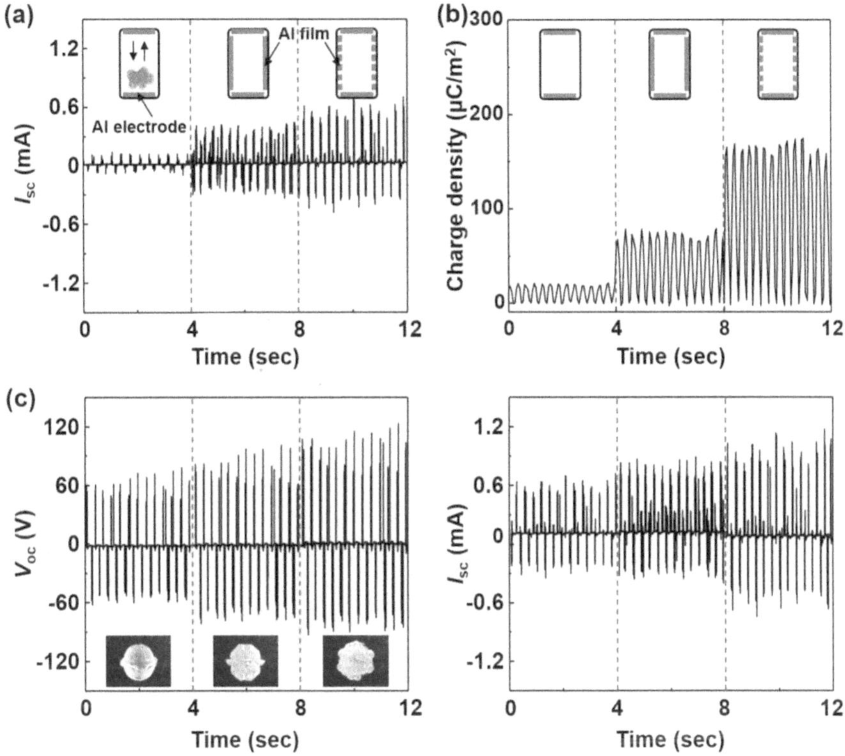

FIGURE 5.15 (a) TENG short circuit currents (I_{sc}) with only Al electrode, Al electrode and Al film without linear pattern, and Al electrode and Al film with linear pattern. (b) TENG charge density. (c) TENGs' open circuit voltages (V_{oc}) and short circuit currents (I_{sc}) with flat, dimpled, and bumpy PDMS balls [25].

Designers of topological optimization have complete voxel freedom. Large discrete numerical models created by finer topological discretizations require plenty of computing resources. Supercomputers have applied giga-voxels to finished structures like airplane wings. Natural convection requires superb topological and physics resolution [24].

Figure 5.15 demonstrates 3D printing. The device optimizes cylinder length and width. A negatively charged 7-mm Al film linear pattern covered the body's interior. Al sheets coated the base part, which served as top and bottom electrodes. By pouring the solution and pulling them from the mold after they set, PDMS balls accepted charges.

A variety of molds controlled ball shape. The body has a fully-packaged cylinder-type TENG and PDMS balls. In practice, shaking arms at 3 Hz and 15 cm amplitude exerts external force (Figure 5.15). The TENG produced 0.12 mA short circuit current (I_{sc}) and 9 V open circuit voltage (Voc) with Al electrodes (Figure 5.16a). Power comes from PDMS balls and Al electrodes transferring charge during contact.

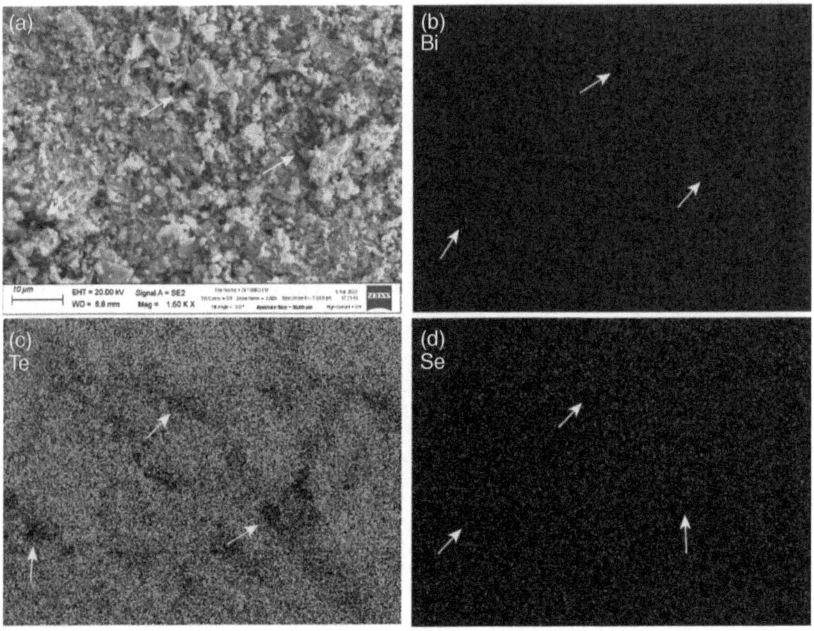

FIGURE 5.16 (a) SEM image and (b–d) EDS elemental maps of n-type samples after sintering [26].

Electrons emitted from the base's two Al electrodes form an external circuit. Due to the limited contact surface, the balls receive less charge, lowering output signals. The electric field formed by two charged particles of the same polarity touching can also release charges. Further negative charge transfer to the balls is needed to boost charge density. The inner surface of Figure 5.15a's inset has an Al layer without the Al electrode to improve contact. Figure 5.16b shows an increase in charge density to 74 μC/m² and output voltage and current to 38 V and 0.33 mA, respectively. PDMS ball charge density increased upon transfer.

5.7 CHARACTERIZATION TECHNIQUES FOR 3D-PRINTED THERMOELECTRIC

Exploring 3D-printed thermoelectrics after printing, TE slurry must dry and be sintered. Pre- and post-sintering SEM pictures of p-type samples are shown in Figure 5.16. Figure 5.17 shows sintering EDS and SEM elemental maps of an n-type sample. No axial pressure is used to print TE slurry. Archimedes estimated a 4.25 g/cm³ sample density. The aggregate-free sample has equal Cu distribution (Figure 5.16a–d). The final material has 1 wt % Cu according to EDX.

With Cu addition, the XRD pattern's Full-width at half maximum (FWHM) decreases and then increases, minimizing at 0.6% Cu. The decrease in FWHM improves crystallinity. Cu addition improves p-type TE material crystallization

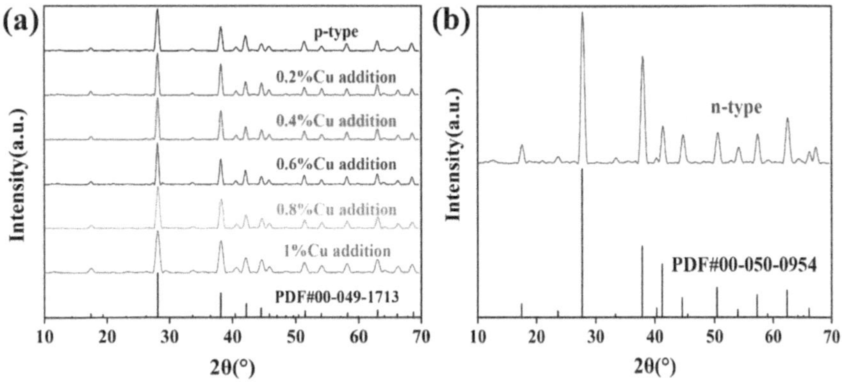

FIGURE 5.17 XRD patterns of 3D-printed TE materials: (a) p-type with different Cu addition content and (b) n-type [26].

and sintering (Figure 5.17a). Figure 5.17b exhibits n-TE XRD. Single-phase $Bi_{0.5}Sb_{1.5}Te_3$ and $Bi_2Se_{0.3}Te_{2.7}$ patterns are seen in p- and n-type samples.

In 3D-printed TE materials, the Seebeck coefficient (α) varies with temperature (Figure 5.18a). In materials without Cu, α decreases with rising temperature for p-type and rises and then decreases for n-type, similar to bulk materials. Cu gradually decreases α in p-type TE materials with increasing Cu content. Additionally, α increases with temperature. The decrease in α may be caused by the higher carrier concentration from adding Cu. Adding Cu to p-type TE materials significantly boosts their electrical conductivity (σ), demonstrating its conductive nature. As Cu concentration increases, the p-type σ rises and then decreases, peaking at 0.6% (Figure 5.18b).

For Cu powder cluster prevention, TE slurry contains Tween-80. The organic solvent Tween-80 decreases TE and Cu in the TE slurry. Tween-80 reduces TE–Cu powder interaction and increases porosity in 3D-printed TE materials after sintering. The p-type σ decreases with over 0.6% Cu content. Cu increases the power factor (PF) in p-type TE materials by affecting α and σ (Figure 5.18c).

With Cu addition, Tween-80 superposes Cu and TE materials in the κ, which increases and then decreases (Figure 5.18d). The study found that Cu enhances ZT initially and ultimately decreases it (Figure 5.18e). The best ZT value for p-type TE with 0.6% Cu is 0.91 at 497 K. N-type ZT_{max} is 0.24 at 432 K. Insufficient conductive chemicals lower ZT. Cu increases ZT_{max}, extending TE device applications and the working temperature range.

For the printing space temperature distribution, a temperature monitoring experiment verified FEM model results. A self-designed temperature measurement apparatus with a PEEK frame and copper wires is shown in Figure 5.19a. Heat transfer can heat or cool copper wire to the ambient air temperature due to its thermal conductivity. Horizontally strung copper wires with a 1.1-mm spacing created the vertical temperature gradient. Figure 5.19b depicts the temperature distribution measuring setup. In the center of the printing space was the

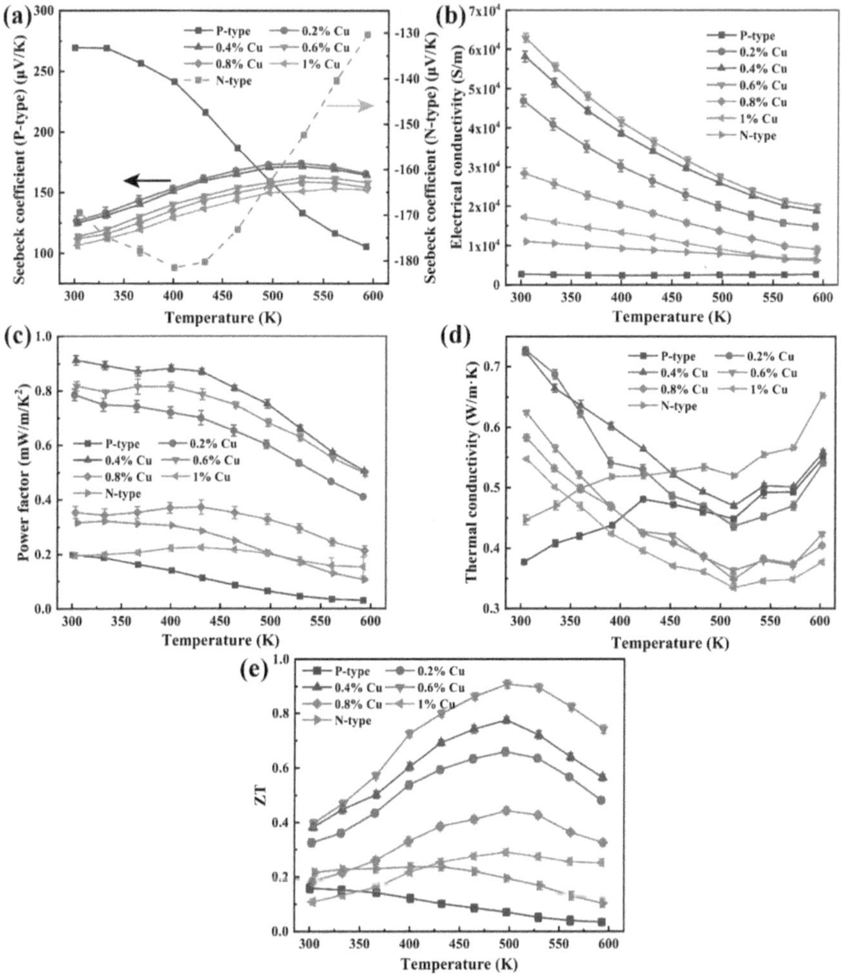

FIGURE 5.18 3D-printed thermoelectric composite's characteristics: (a) Seebeck coefficient, (b) electrical conductivity, (c) power factor, (d) thermal conductivity, and (e) ZT value [26].

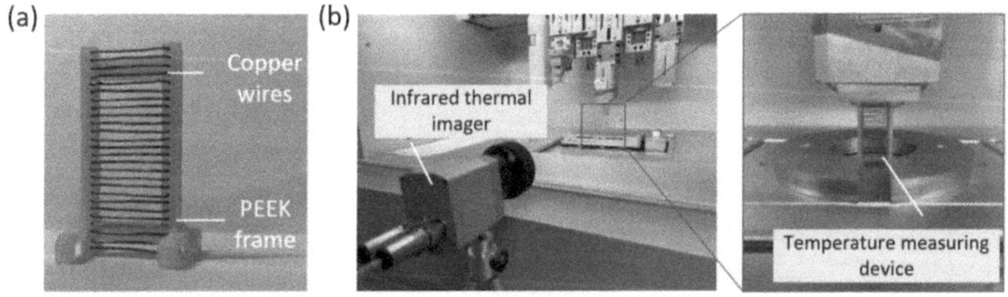

FIGURE 5.19 Printing space temperature measurement: (a) self-made temperature sensor and (b) experiment for measuring temperature distribution [27].

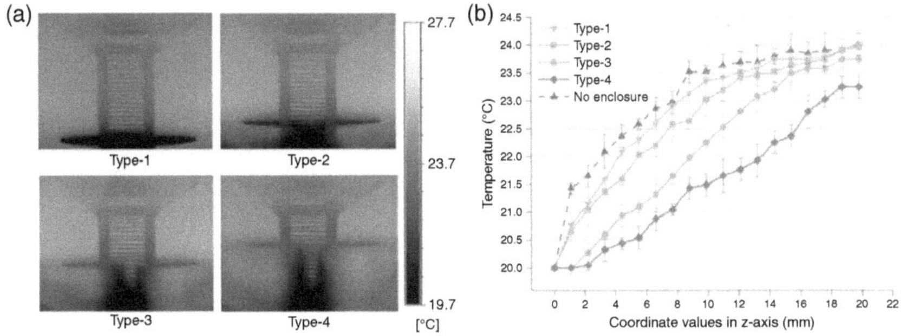

FIGURE 5.20 Measured temperature distribution: (a) use thermal pictures for printing and (b) temperature curves from the experiment [27].

self-designed gadget on the two-axis linear stage. To measure temperature accurately, it is important to avoid touching the cooled plate's top. The multi-layered enclosures were gapped to avoid obscuring an infrared thermal imager from measuring the organized copper wire temperatures.

Figure 5.20a shows temperature measuring experiment infrared thermal images. Figure 5.20b shows the temperature curves and data for each experiment without the enclosure. Thus, each copper wire is hotter than its neighboring space, and the enclosure height affects both numerical and experimental data [27].

TE properties and Equation 1 zT values for pre- and post-annealing cast samples are shown in Figure 5.21a–c and d. Increasing temperature decreases thermal conductivity until ~600 K, where bipolar contributions enhance it (Figure 5.22a). With temperature, pre- and post-annealing electrical resistivities decrease monotonically, exhibiting semi-conductivity (Figure 5.22b). Both states have negative Seebeck coefficients (Figure 5.22c), with electrons as the main charge carriers (n-type). Minority carrier (hole) contributions decrease the absolute Seebeck coefficient and cause the parabolic form at higher temperatures during band gap excitation (Figure 5.22d).

In Figure 5.22, single pn leg output power simulations at varying copper electrode thicknesses are displayed. As seen in Figure 5.22, output power rapidly increases and stabilizes with copper electrode thickness. Power production peaked at 6.02 mW at a 0.23 mm copper electrode thickness. Thermal generator power depends on current density. Electron and hole mobility limit thermoelectric element current density. High current density follows charge mobility, although high carrier density compensates for low-mobility semiconductors.

Examination of three flexible substrate layers with varied heat conductivity was conducted. Adhesive attached all 0.2 mm flexible substrate layers to the FTEG. Figure 5.23a displays flexible FTEG pictures. A K-type thermocouple measured steady-state temperatures of 147°C and 82°C for the hot and cold sides in the experiment. The association is shown in Figure 5.23b. FTEG (SR), SR@G, and the graphite sheet had steady-state V_{oc} of 0.75, 1.0, and 1.1 V, respectively.

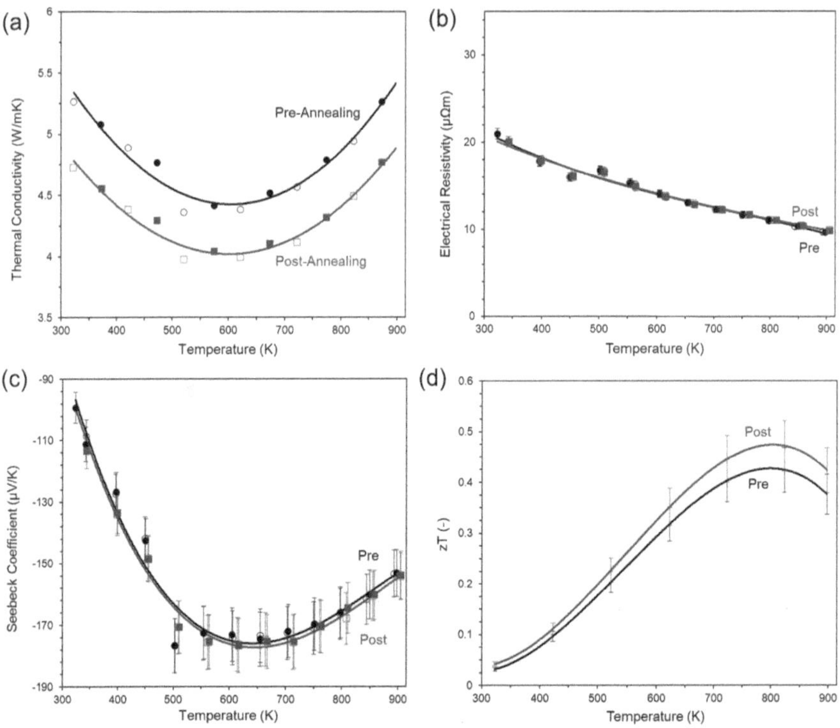

FIGURE 5.21 Analysis of cast TiNiSn plate thermoelectric properties before and after one week of 850°C annealing: (a) thermal conductivity, (b) electrical resistivity, (c) Seebeck coefficient, and (d) figure of merit, zT. Solid lines calculate zT using polynomial best-fit curves, while filled and open symbols show heating and cooling data. Error bars are smaller than thermal conductivity data [28].

The internal resistance and the effect of the flexible substrate on FTEG output power were measured with load resistors on the device's output terminal.

In Figure 5.23c, load resistor settings affect output power. At 65°C and a 2 Ω load resistor, the FTEG and graphite sheet produce the greatest output power (160 mW). This investigation's 5×5 cm² FTEG module prototype has an electrical resistance of approximately 2 Ω. Figure 5.23d presents power density-temperature difference relationships.

The finite element method (FEM) is used to generate a steady-state model and analyze design parameters to assess ΔT1. The simulation maintains 147°C and 82°C for the hot and cold sides. Conduction transfers heat from the hot to the cold side of the FTEG. The simulations reveal that the bendable substrate's low thermal conductivity affects heat dissipation and the thermoelectric components' hot and cold junction temperature difference.

With its high conductivity and low-temperature resistance, the tested $KM_{15}G_8H_{0.5}Ca_5$ gel's tensile deformation resistance was evaluated. Figure 5.24a shows that the $KM_{15}G_8H_{0.5}Ca_5$ gel GF grew with tensile strain to 4.40 at 400%.

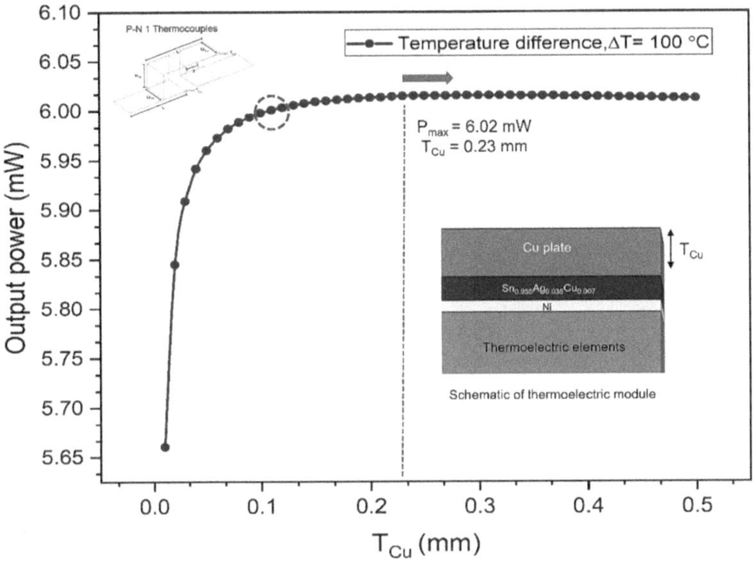

FIGURE 5.22 Power of thermocouples depends on copper contact thickness [29].

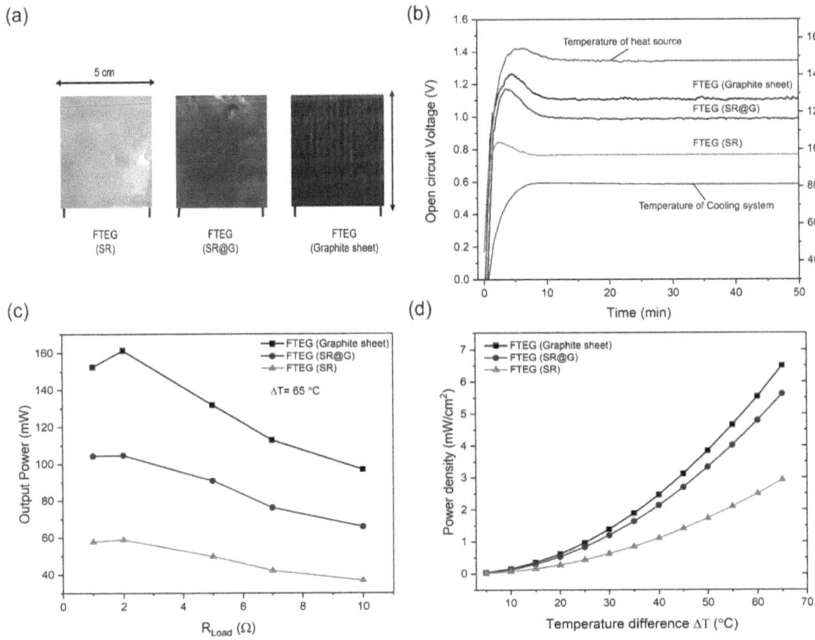

FIGURE 5.23 (a) Photographs depicting the graphite sheet, SR@G, SR, and device dimensions; (b) monitoring the temperature of an open circuit using flexible substrates; (c) power of FTEG module output; and (d) power density and heat spreader materials at 65°C temperature differences [29].

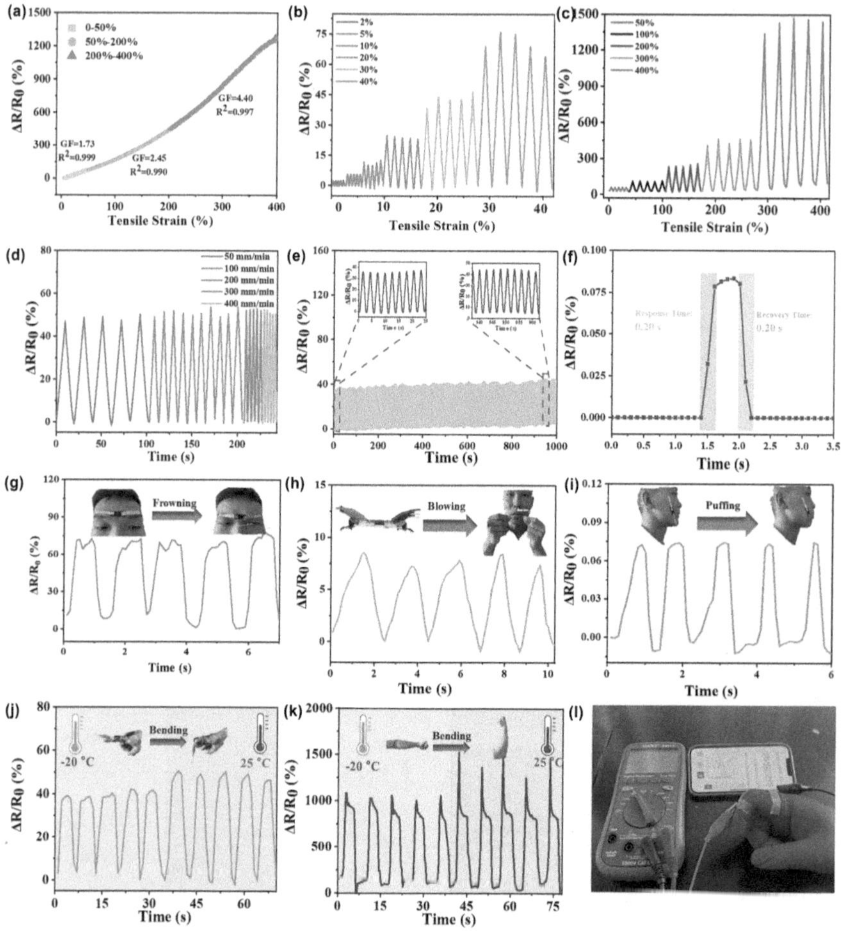

FIGURE 5.24 (a) $KM_{15}G_8H_{0.5}Ca_5$ gel sensor strain-relative resistance curve. Under dynamic strain loading, sensor relative resistance changes: (b) small strain (2%–40%) and (c) large strain (50%–400%). (d) Comparing $KM_{15}G_8H_{0.5}Ca_5$ gel sensor resistance under 20% strain cycle loading and unloading at various tensile rates. (e) Cyclic loading-unloading of $KM_{15}G_8H_{0.5}Ca_5$ gel-based sensor with 20% tensile strain results in relative resistance variations. (f) Time of response/recovery for $KM_{15}G_8H_{0.5}Ca_5$ gel sensor. The relative resistance change curve of the different body parts on the (g) forehead, (h) mouth, and (i) cheek. The sensor responds to joint motion at –20°C and 25°C: (j) finger and (k) elbow. (l) Digital image of the sensor with wireless detection [30].

Figure 5.24b and c reveals that the $KM_{15}G_8H_{0.5}Ca_5$ gel-based sensor was highly sensitive to small-scale (2%–40%) and large-scale (50%–400%) strain (Figure 5.24c and d). Under 20% tensile strain throughout long-term tensile cycles, 300 stretching-releasing cycle trials showed stability and durability (Figure 5.24e).

Strain and release reaction recovery occurred in 200 ms (Figure 5.24f). Continuous gel stretching, trace water evaporation, and polymer chain growth per unit volume cause the relative resistance curve to change. The $KM_{15}G_8H_{0.5}Ca_5$ gel has a higher GF and conductivity (2.07 S/m) than other hydrogels.

Frowning, blowing, and puffing changed relative resistance (Figure 5.24g–i). Thus, frowning, blowing, and puffing changed resistance by 70%, 10%, and 0.07%, respectively. Expression changes stretched the hydrogel, increasing electrons and electron transport distance, which altered the resistance. Human elbow, knee, and breathing movements may be detected by the hydrogel. The $KM_{15}G_8H_{0.5}Ca_5$ gel-based sensor demonstrated excellent reproducibility under finger and elbow bending at low and room temperatures (Figure 5.24j and k). Using the relative resistivity rate of change, hydrogel strain sensors can distinguish multiple human motions (Figure 5.24l).

5.8 POTENTIAL OF 3D PRINTING FOR THE FABRICATION OF THERMOELECTRIC MATERIALS AND DEVICES

Green energy conversion technology allows waste heat to be used to cool spots with thermoelectric devices. However, most thermoelectric devices use toxic and expensive materials, which restrict their use. High-temperature fabrication limits the compatibility of biocompatible, flexible substrates. Many biocompatible and conformable devices are made using printed electronics. Customizing printed thermoelectric devices for their use is more affordable. Polymers and other bio-friendly materials have been intensively explored for low-cost, lightweight, and flexible thermoelectric devices. Low-temperature printing can be used to print these materials on biocompatible surfaces.

Performance-opposing characteristics are optimized in thermoelectricity. Printing thermoelectric materials requires ink and optimization. Researchers have examined thermoelectric materials, substrates, and TEG efficiency [31].

Direct-writing 3D printing was used to create annular and in-plane TEs. With sophisticated equipment, the 3D-printed devices' output voltage and power were evaluated at different temperatures. The annular device's output was tested with a tube-heated rod. Finally, the device was inserted into a metal ingot. A heat sink and ingots slowed the temperature rise outside the device. Filling the heating rod-inner pipe wall and device-metal ingot through-hole gaps with heat-conducting silicone grease reduced heat loss. FEM simulated the annular TE device output to confirm the experiments.

Direct-writing 3D printing of TE composites with good TE properties and controlled shapes allowed for the flexible structure design of TE devices to fit curved heat sources and reduce heat loss. Direct-writing 3D printing was used to create annular and in-plane TEs. This study used a low-k glass substrate for the in-plane TE device. Figure 5.25a shows silver paste joining p- and n-type TE legs sequentially. The TE legs must be long enough to accommodate a large temperature differential and have a high integration degree to build a suitable micro-device with higher output voltage and power. In contrast, TE devices should have low

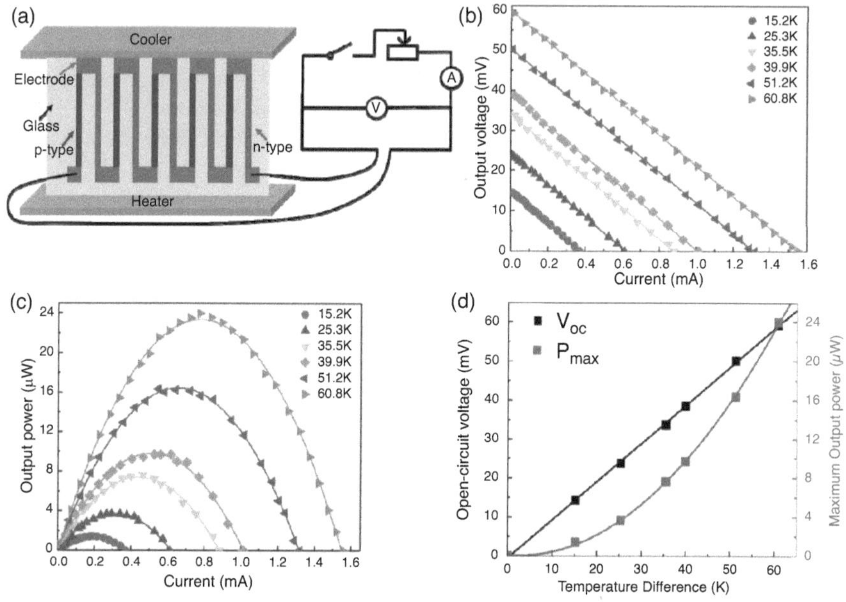

FIGURE 5.25 (a) In-plane device schematic with five pairs of TE legs. (b and c) Current-dependent output voltage and power curves of the TE device for 9.8 to 39.9 K temperature differences. (d) Temperature variation affects TE device open-circuit voltage and maximum output power [32].

internal resistance to minimize leg length and diameter. Low metal electrode-TE leg contact resistance was also required.

Five pairs of 800 mm diameter, 10 mm TE legs were fabricated on the glass substrate, considering these two parameters. The TE legs were joined in series using silver paste as an electrode after 6 h of heat treatment. Tests of the device's generation performance employed custom equipment. Between two copper billets, the TE device had a digital collector and a resistance box in series. Figure 5.25b–d shows the TE device's output voltage and power curves as a function of current from 15.2 to 60.8 K.

At certain temperatures, the output voltage and current were linear, while power and current were parabolic. The open-circuit voltage was 59.21 mV, the short-circuit current was 1.56 mA, and the maximum output power was 23.97 mW at 60.8 K. Lower than the absolute S values of the n- and p-type materials, the TE legs had an average S value of 194.77 mV/K. Silver paste filled a certain volume, so the two ends of the TE legs were not in total contact with the device's hot and cold ends; therefore, their temperature differential was smaller than that of the thermocouples.

TE materials had a higher S value than the device. From the voltage–current curve slope, the TE device's resistance was 37.93 Ω, which was far greater than that of the TE legs. Silver paste-TE leg contact resistance caused this. A better electrode material could lower the contact resistance and enhance the device's output.

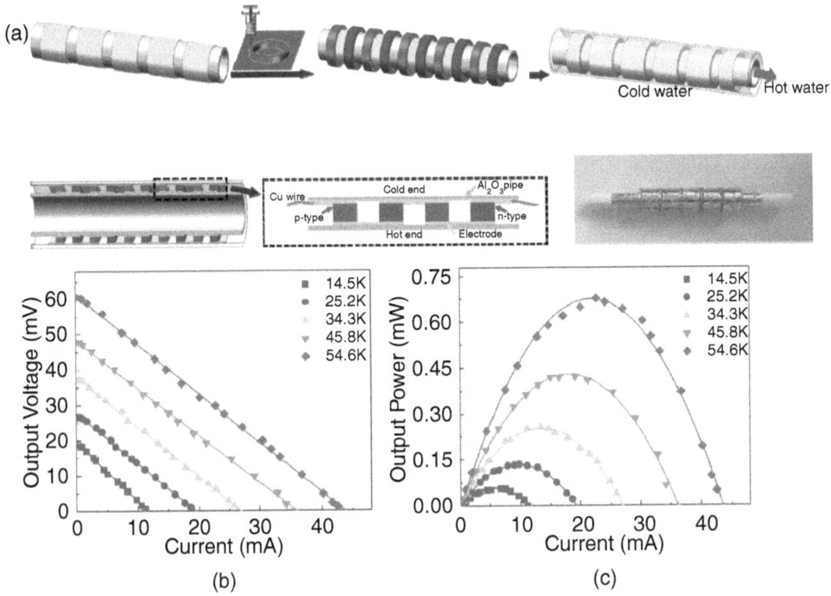

FIGURE 5.26 (a) Photograph and schematic of annular device preparation with five pairs of TE legs. (b and c) Output voltage and power of annular devices as functions of current at 14.5–54.6 K temperature difference [32].

The maximum output power ranged from 1.42 to 23.97 mW, and the open-circuit voltage ranged from 14.38 to 59.21 mV for temperature fluctuations between 15.2 K and 60.8 K. Open-circuit voltage and temperature difference were linear, whereas maximum output power and temperature difference were parabolic, matching the TE device's output characteristics. Direct-write 3D printing creates TE composites for TE devices.

Then, direct-writing 3D printing created advanced gadgets that older technologies could not. 3D-printed annular TE devices were developed. Two hollow aluminum oxide tubes supported the annular TE device. A low-k silica gel ring separated the annular p- and n-type TE legs. The annular TE device works effectively with non-planar or fluid-medium cold and heat sources that flow heat along the cylindrical heat source. Figure 5.26a depicts the annular device (five pairs of TE legs) preparation process in both photo and schematic form.

The complete TE gadget was sandwiched between two hollow pipes inserted into its exterior. The temperature difference between the hot and cold water in the inner and outer pipes can power the TE devices. This method could generate electricity from nuclear power plant waste hot water or gas. The TE annulus thermal stress is low due to its small linear expansion coefficient and gentle temperature gradient. The electrode-TE annulus connection and homogeneous thermal effect reduce thermal stress in the device.

For temperature fluctuations ranging from 14.5 to 54.6 K, the annular device's output voltage and power are current-dependent, as shown in Figure 5.26b and c.

The device had a 60.80 mV open-circuit voltage, a 42.51 mA short-circuit current, and 0.68 mW output power at 54.6 K. The annular device's voltage-current curve slope estimated its resistance to be 1.43 Ω, and the two TE legs had an average S value of 223 mV/K. Because they could manage the largest temperature difference between the hot and cold sides, our in-plane and annular devices had the highest open-circuit voltage per thermocouple.

Using H2 for 2 h may remove particle surface oxide coatings, improving electrical properties (Figure 5.27a). Per XRD patterns (Figure 5.27b), all samples were fully sintered without secondary phases. Sb concentration caused XRD peak shifts to larger angles, indicating 3D printing generated single-phase composition-engineered BiSbTe.

The lattice of $Bi_xSb_{2-x}Te_3$ is stretched due to the larger atomic radius of Bi (1.60 Å) compared to Sb (1.45 Å). SEM images of the samples (Figure 5.27c) show well-sintered grains of tens of micrometers and pores of several micrometers. The sintering effect of Sb_2Te_4 2– chalcogenidometalate (ChaM) led to significant volume shrinkage in printed products after heating. Sintering shrinkage was consistent regardless of composition. Figure 5.27d shows 75%–80% densities and 20%–25% porosities in all samples. ChaM binder content affected sintering shrinkage but not particle composition.

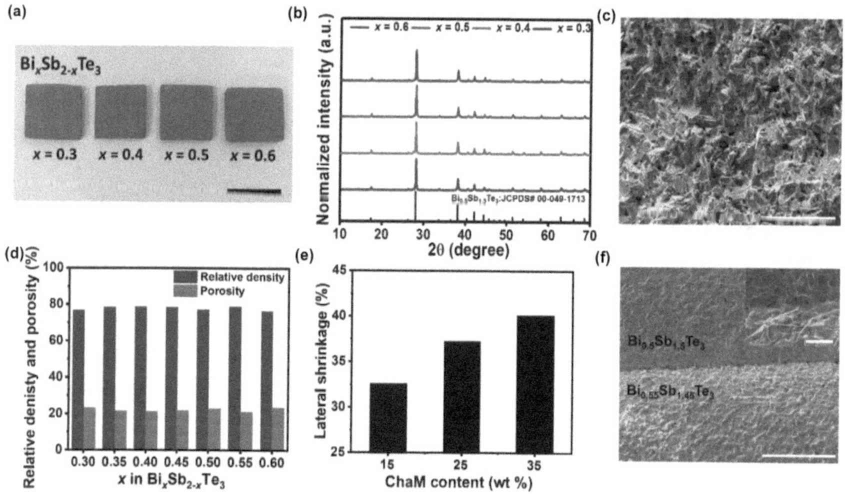

FIGURE 5.27 3D-printed segmented TE materials. (a) Photograph and (b) XRD patterns of 3D-printed structures utilizing TE inks with $Bi_xSb_{2-x}Te_3$ particles (x=0.3–0.6). The vertical lines show bulk $B_{i0.5}Sb_{1.5}Te_3$ peaks (JCPDS: 00-049-1713). (c) SEM of $Bi_{0.5}Sb_{1.5}Te_3$-infused TE ink-printed 3D materials. (d) Relative density and porosity of 3D-printed structures employing TE inks with $Bi_xSb_{2-x}Te_3$ particles (x=0.3–0.6). (e) After heat treatment, 3D-printed Bi0.5Sb1.5Te3 blocks shrink laterally depending on ChaM concentration. (f) SEM image of 3D-printed segmented bi-block TE material $Bi_{0.55}b_{1.5}Te_3$/ChaM 25% and $Bi_{0.55}Sb_{1.55}Te_3$/ChaM25% interface. Inset: high-resolution SEM picture [33].

ChaM from 15% to 35% increased lateral shrinkage from 32% to 40% (Figure 5.27e). This shrinkage rate mismatch causes interface cracks in segmented samples with different ChaM concentrations. Sintering fractured surfaces between printed objects with 15% and 35% ChaM binder. For sequential printing of multi-material segments with $Bi_{0.55}Sb_{1.45}Te_3$, $Bi_{0.55}Sb_{1.5}Te_3$, and $Bi_{0.35}Sb_{1.65}Te_3$, 25% ChaM inks are utilized. SEM shows well-fused segment interfaces due to ChaM's sintering assistance promotion (Figure 5.27f). These results show that current 3D printing technology can cost-effectively manufacture high-performance segmented TE material, which needs difficult multistep operations including synthesizing, dicing, and bonding.

REFERENCES

1. Oztan, C., Ballikaya, S., Ozgun, U., Karkkainen, R., & Celik, E. (2019). Additive manufacturing of thermoelectric materials via fused filament fabrication. *Applied Materials Today, 15*, 77–82.
2. Tzounis, L., & Bangeas, P. (2022). 3D printing and nanotechnology. In Papadopoulos, V. N., Tsioukas V., Suri, J. S. (eds.). *3D Printing: Applications in Medicine and Surgery* (Vol. 2, pp. 7–26). Elsevier.
3. Wu, Y., Ma, Y., Zheng, H., & Ramakrishna, S. (2021). Piezoelectric materials for flexible and wearable electronics: a review. *Materials & Design, 211*, 110164.
4. Shakeel, M., Rehman, K., Ahmad, S., Amin, M., Iqbal, N., & Khan, A. (2021). A low-cost printed organic thermoelectric generator for low-temperature energy harvesting. *Renewable Energy, 167*, 853–860.
5. Wu, Y., An, C., & Guo, Y. (2023). 3D printed graphene and graphene/polymer composites for multifunctional applications. *Materials, 16*(16), 5681.
6. Bian, M., Xu, Z., Tang, X., Jia, H., Wang, Y., & Cabot, A. (2024). Solution-based 3D printed thermoelectric composite for custom-shaped thermoelectric generator applications. *Journal of Alloys and Compounds, 976*, 173202.
7. Su, N., Zhu, P., Pan, Y., Li, F., & Li, B. (2020). 3D-printing of shape-controllable thermoelectric devices with enhanced output performance. *Energy, 195*, 116892.
8. Siddique, A. R. M., Rabari, R., Mahmud, S., & Van Heyst, B. (2016). Thermal energy harvesting from the human body using flexible thermoelectric generator (FTEG) fabricated by a dispenser printing technique. *Energy, 115*, 1081–1091.
9. Yang, S. E., Kim, F., Ejaz, F., Lee, G. S., Ju, H., Choo, S., ... & Son, J. S. (2021). Composition-segmented BiSbTe thermoelectric generator fabricated by multimaterial 3D printing. *Nano Energy, 81*, 105638.
10. Jia, N., Cao, J., Tan, X. Y., Dong, J., Liu, H., Tan, C. K. I., ... & Suwardi, A. (2021). Thermoelectric materials and transport physics. *Materials Today Physics, 21*, 100519.
11. Nayak, R., Shetty, P., Selvakumar, M., Rao, A., & Rao, K. M. (2022). Formulation of new screen printable PANI and PANI/Graphite based inks: printing and characterization of flexible thermoelectric generators. *Energy, 238*, 121680.
12. Nandihalli, N., Liu, C. J., & Mori, T. (2020). Polymer based thermoelectric nanocomposite materials and devices: fabrication and characteristics. *Nano Energy, 78*, 105186.
13. Orrill, M., & LeBlanc, S. (2017). Printed thermoelectric materials and devices: fabrication techniques, advantages, and challenges. *Journal of Applied Polymer Science, 134*(3), 44256.

14. Zhou, X., Wu, L., & Wang, J. (2023). Recent developments in conductive polymer composites for fused deposition modeling. *Composites Part A: Applied Science and Manufacturing, 174*, 107739.
15. Candere, A. C., Miché, N., Bernagozzi, M., Saglam, M., Georgoulas, A., Aydin, O., & Marengo, M. (2024). A novel fabrication method for polymeric flat plate pulsating heat pipe via additive manufacturing. *Applied Thermal Engineering, 241*, 122398.
16. Wu, Y., Sun, K., Yu, S., & Zuo, L. (2021). Modeling the selective laser melting-based additive manufacturing of thermoelectric powders. *Additive Manufacturing, 37*, 101666.
17. Du, Y., Chen, J., Meng, Q., Dou, Y., Xu, J., & Shen, S. Z. (2020). Thermoelectric materials and devices fabricated by additive manufacturing. *Vacuum, 178*, 109384.
18. Bian, M., Xu, Z., Meng, C., Zhao, H., & Tang, X. (2022). Novel geometric design of thermoelectric leg based on 3D printing for radioisotope thermoelectric generator. *Applied Thermal Engineering, 212*, 118514.
19. Wu, Y., Yang, J., Chen, S., & Zuo, L. (2018). Thermo-element geometry optimization for high thermoelectric efficiency. *Energy, 147*, 672–680.
20. Eom, Y., Kim, F., Yang, S. E., Son, J. S., & Chae, H. G. (2019). Rheological design of 3D printable all-inorganic inks using BiSbTe-based thermoelectric materials. *Journal of Rheology, 63*(2), 291–304.
21. Sobianin, I., Psoma, S. D., & Tourlidakis, A. (2024). A 3D-printed piezoelectric microdevice for human energy harvesting for wearable biosensors. *Micromachines, 15*(1), 118.
22. Rösch, A. G., Franke, L., Mallick, M. M., & Lemmer, U. (2023). Optimizing printed thermoelectric generators with geometry and processibility limitations. *Energy Conversion and Management, 279*, 116776.
23. Yuan, J., & Zhu, R. (2020). A fully self-powered wearable monitoring system with systematically optimized flexible thermoelectric generator. *Applied Energy, 271*, 115250.
24. Lazarov, B. S., Sigmund, O., Meyer, K. E., & Alexandersen, J. (2018). Experimental validation of additively manufactured optimized shapes for passive cooling. *Applied Energy, 226*, 330–339.
25. Lee, J. P., Ye, B. U., Kim, K. N., Lee, J. W., Choi, W. J., & Baik, J. M. (2017). 3D printed noise-cancelling triboelectric nanogenerator. *Nano Energy, 38*, 377–384.
26. Bian, M., Xu, Z., Tang, X., Jia, H., Wang, Y., & Cabot, A. (2024). Solution-based 3D printed thermoelectric composite for custom-shaped thermoelectric generator applications. *Journal of Alloys and Compounds, 976*, 173202.
27. Yu, K., Gao, Q., Lin, Z., Wang, F., Zhang, P., Liu, Y., & Lu, L. (2024). Investigation of the temperature gradient control in the printing space for the material extrusion of medical biodegradable hydrogel. *Computers in Biology and Medicine, 168*, 107722.
28. Pröschel, A., Lau, M. L., Guillen, D. P., & Dunand, D. C. (2024). Combining direct ink writing with reactive melt infiltration to create architectured thermoelectric legs. *Chemical Engineering Journal, 479*, 147845.
29. Gobpant, J., Klongratog, B., Rudradawong, C., Sakdanuphab, R., Junlabhut, P., Nuthongkum, P.,& Sakulkalavek, A. (2024). High-performance flexible thermoelectric generator based on silicone rubber and cover with graphite sheet. *Applied Thermal Engineering, 236*, 121656.
30. Hou, M., Yu, M., Liu, W., Zhang, H., Wang, Z., Du, J., ... & Xu, J. (2024). Mxene hybrid conductive hydrogels with mechanical flexibility, frost-resistance, photothermoelectric conversion characteristics and their multiple applications in sensing. *Chemical Engineering Journal, 483*, 149299.

31. Hossain, M. S., Li, T., Yu, Y., Yong, J., Bahk, J. H., & Skafidas, E. (2020). Recent advances in printable thermoelectric devices: materials, printing techniques, and applications. *RSC Advances*, *10*(14), 8421–8434.
32. Su, N., Zhu, P., Pan, Y., Li, F., & Li, B. (2020). 3D-printing of shape-controllable thermoelectric devices with enhanced output performance. *Energy*, *195*, 116892.
33. Yang, S. E., Kim, F., Ejaz, F., Lee, G. S., Ju, H., Choo, S., ... & Son, J. S. (2021). Composition-segmented BiSbTe thermoelectric generator fabricated by multimaterial 3D printing. *Nano Energy*, *81*, 105638.

6 Efficiency and Performance of Thermoelectric Materials and Devices

6.1 INTRODUCTION

Waste heat is converted into electricity by thermoelectric (TE) materials, improving fuel economy and providing a viable alternative energy source for innovative energy solutions. Many TE materials are needed for high-performance devices. High-performance TE devices aid TE material research and application. Over the past two decades, more good TE materials than Bi_2Te_3, PbTe, and SiGe-based materials have been discovered and produced. High-performance TE materials' mechanisms were discovered theoretically and experimentally. Due to many strategies that make it difficult to enhance every parameter without affecting efficiency, significant efforts are still needed to bring the ZT near the theoretical limit [1,2]. Most high-ZT materials should be thermoelectrically efficient. Heated thermoelectric engines generate power. Many high-ZT materials increase thermoelectric efficiency [3].

6.2 EFFICIENCY AND PERFORMANCE ENHANCEMENT STRATEGIES FOR THERMOELECTRIC MATERIALS AND DEVICES

Along with nanostructuring to lower heat conductivity, this unique physical method could improve zT and enable more thermoelectric devices. Thermal production depends on material efficiency. Due to inefficient thermoelectric materials that stay near 10% of Carnot efficiency for long periods, reliable and efficient ones have not yet been identified for large-scale commercial use. ZT must be large, even though the three transport coefficients are not independent. To optimize power factor ($PF = S^2\sigma$), narrow-band semiconductors with a carrier density of 10^{19}–$10^{21}/cm^3$ are often employed. Strongly scattering lattice phonons lower heat conductivity. Band, phonon, defect, and interface engineering improved ZT for thermoelectric materials [4].

An effective thermoelectric module uses n- and p-type parts. Compared to Mg_2IV-based materials, p-type higher manganese silicide (HMS) has higher ZT values and better mechanical properties. MnSi intermetallic layered precipitates

 DOI: 10.1201/9781003639343-6

were found in Czochralski and Bridgman single crystalline specimens in early HMS studies. Chemical vapor transfer produced pure MnSi-free HMS single crystals. HMS precipitated the unstable MnSi phase during consolidation due to its instability at high temperatures. The cost and mechanical qualities of polycrystalline HMS are superior to single crystal growth. Doping HMS with Al, Ge, Cr, Re, Ru, W, and others improves p-type TE. To optimize hole density and lower lattice thermal conductivity, doping increases point defect phonon dispersion. Single, double, or sophisticated multi-doping can improve ZT values to near unity, but the best carrier concentration is unclear. Microstructure optimization, electronic band engineering, and multi-doping may increase HMS ZT [5].

Many TEG materials are commercially used. This material has three operational temperature classes [6]:

1. Low-temperature materials (450 K): bismuth (Bi) alloys with antimony, tellurium, or selenium;
2. Lead alloys for intermediate temperatures (850 K); and
3. Materials manufactured from SiGe alloys can endure 1300 K.

PbTe, GeTe, and SiGe sublimation at 350 K to 1150 K for TE protective coatings was studied using DSC and TGA. PbTe and GeTe show significant thermal property changes and weight loss at working temperatures. SiGe had a 0.0001% weight difference and no thermal impacts up to 1200 K. Materials, forming methods, and protective coatings for multisection TEs exist. Plasma-enhanced Si_3N_4 and SiO_2 protective layers worked and stayed stable at TE temperatures. Si_3N_4 reduced PbTe and GeTe TEM weight by 0.18% and 0.14%. TEM efficiency and operating temperatures must increase to optimize TEG efficiency. Two methods improve TEG temperature range. Predetermined temperatures limit each TEM TE segment's Z. Alternate TEM makes thermoelectric modules, solving multisectional TE design and technology issues.

Calculate and optimize TE design using TEM thermal, electrical, and operation temperatures. Each thermoelement segment is sized by the temperature profile where its TEM has the highest thermoelectric figure of merit. A multisectional TEM design was efficiently built using this technology. Contact systems for TE have been created, and material choices justified. Contact solutions with low resistance (10^{-9} $\Omega.m^2$), high adhesive strength (22.0 MPa), and stable temperatures were created. Variable melting point soldering, high melting point eutectic compounds, and gold commutation layers were used for TE section commutation. TEM's TCLE investigation replicates TE's linear dimensions across temperatures. Patented composite TE-containing contact systems reduce thermal expansion stress. At high temperatures, plasma-assisted deposition and Si_3N_4 and SiO_2 thin-film protective coatings on TE legs may prevent TEM thermoelement sublimation [7].

Bismuth telluride/skutterudite and high-entropy-stabilized chalcogenide segmented modules have high conversion efficiencies (12%) due to thermoelectric optimization and rational device design. Segmented device reliability and stability must be addressed. The use and evaluation of thermoelectric devices require dependability, stability, and efficiency. Thermoelectric devices need structural stability, material deterioration resistance, and matching component thermal expansion coefficients. Different-material segmented thermoelectric devices have constraints.

1. Device instability can result from thermal cycling interface failure in thermoelectric materials with differing thermal expansion coefficients.
2. Using barrier layers to inhibit element migration between thermoelectric materials is common.

These interfacial barrier materials may reduce energy loss and conversion efficiency due to higher electrical and thermal resistances. Complex soldering is needed to make thermoelectric connections in segmented materials. Thermoelectric material sintering temperatures prevent one-step sintering, complicating manufacturing [8].

Many approaches improve TE material characteristics. To improve total TE efficiency without increasing zT, adjusting one parameter may compensate. In four main ways, decoupling electronic and phonon-dependent transport characteristics has considerably influenced transport parameters:

1. Nanostructuring,
2. Alloying/chemical doping,
3. Phase engineering, and
4. Defect creation.

Electronic and phonon transport is better in semiconductors than in semimetals and metals due to charge carrier density and dispersion curves. Adjusting TE parameters does not affect other material properties. Bi_2Te_3, PbSe, and other semiconductors have narrower electronic band gaps than metals or semimetals. Lower carrier effective masses from band convergence and nestification promote degeneracy. They disconnected the crucial electronic-thermal transfer link [9].

6.2.1 Nanostructured Thermoelectric Materials

Recent nanotechnology advances have improved TE materials' thermoelectric properties. Raising Z decreases the lattice component and lowers bulk nanostructured materials' thermal conductivity due to nanograin and nanoparticle phonon scattering. Compressing bulk nanostructured TE materials to a density above 95% avoids orders of magnitude electrical conductivity variations induced by a few percent density change. Due to charge carrier scattering, less dense materials conduct less electricity. By lowering heating time, spark plasma sintering avoids grain growth over hot pressing. Useful nanostructured TE materials have good Z values, thermal stability, and mechanical qualities. Thermal cycling allows TEGs to work under critical conditions at many temperatures for long periods.

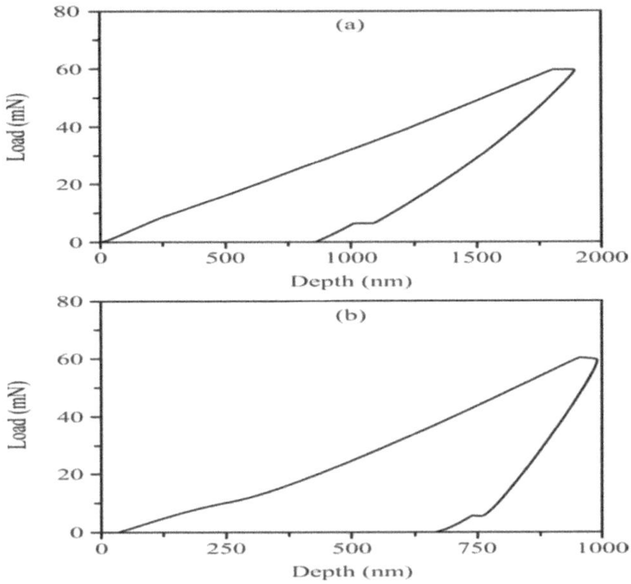

FIGURE 6.1 Nanostructured load-unloading curves: (a) PbTe-40 and (b) GeTe-60 [10].

Under-100-nm grains make up less than 50%. More indenter penetration occurs from 20 to 60 mN stress. Figure 6.1 displays the nanostructured PbTe40 and GeTe-40 loading-unloading trends at 60 mN [10].

$Bi_2Te_{3-x}Se_x$ P- and n-type sources cool room-temperature SSEs. TEs are inefficient because p-type $Bi_{2-x}Sb_xTe_3$ dominates n-type $Bi_2Te_{3-x}Se_x$ near ambient temperature. $Bi_{2-x}Sb_xTe_3$ TE performance is shown in Figure 6.2. It yields good p-type TEs solely, unlike n-type $Bi_2Te_{3-x}Se_x$. The following mechanisms collaborate to achieve success:

1. The donor-like effect causes Bi_2Te_3 to have too many condition electrons.
2. Bi_2Te_3 and $Bi_2Te_{3-x}Se_x$ enhance conduction band degeneracy. TE modules with similar chemical compositions of n- and p-legs are stronger and more durable [11].

The κT change of pure and Sb-substituted Cu_2Te is temperature-dependent (Figure 6.3). Compared to the pure sample, $Cu_{1.95}Sb_{0.05}Te$ and $Cu_{1.90}Sb_{0.1}Te$ samples have lower k_T values up to 523 K and slightly higher values at 623 K. Anharmonicity and increased phonon-phonon scattering occur in low κT in $Cu_{1.95}Sb_{0.05}Te$ and $Cu_{1.90}Sb_{0.1}Te$ samples due to the higher Sb ionic radius at the Cu site. $Cu_{1\cdot8}Sb_{0\cdot2}Te$ exhibits a higher κT rate than other samples, possibly due to the Sb_2Te_3 secondary phase.

Figures 6.4 and 6.5 provide κe and κL values for $Cu_{2-x}Sb_xTe$ (x=0.05, 0.1, 0.2). In the $Cu_{1\cdot95}Sb_{0\cdot05}Te$ sample, lower ρ values lead to greater κe values than the pure sample. The $Cu_{1\cdot95}Sb_{0\cdot05}Te$ sample has a lower κL than the pure sample due

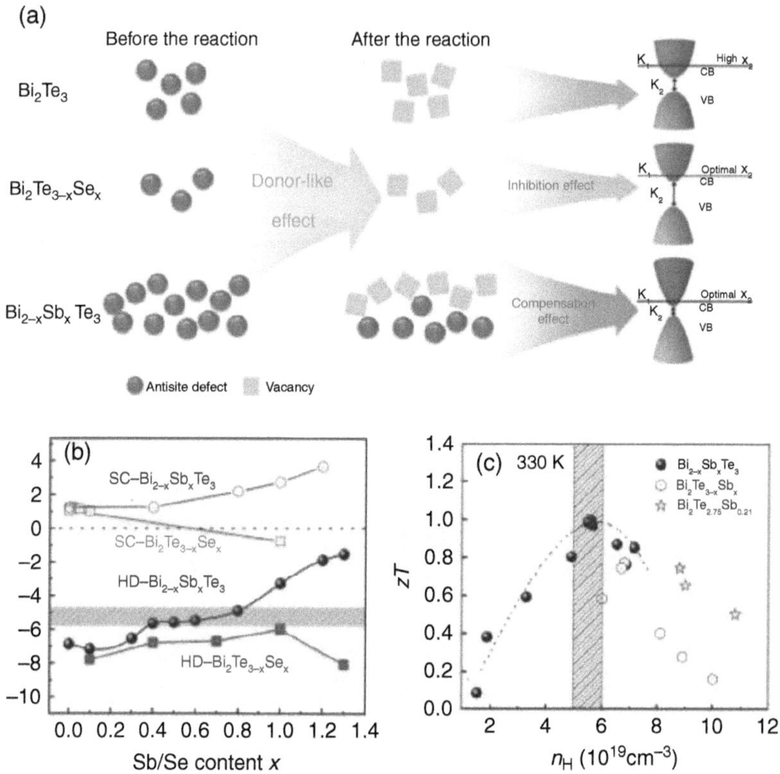

FIGURE 6.2 $Bi_{2-x}Sb_xTe_3$ thermoelectric capability: (a) schematic of intrinsic point defect variation, (b) comparison of room-temperature carrier concentration, and (c) carrier concentration-dependent figure of merit at 330 K [11].

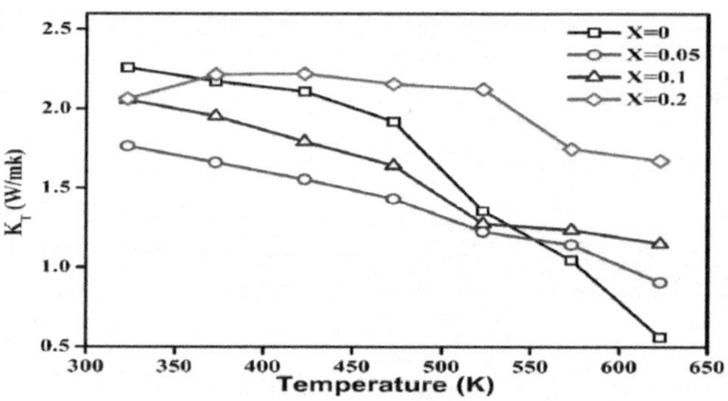

FIGURE 6.3 Temperature-dependent total thermal conductivity of $Cu_{2-x}Sb_xTe$ (x=0, 0.05, 0.1, 0.2) samples [12].

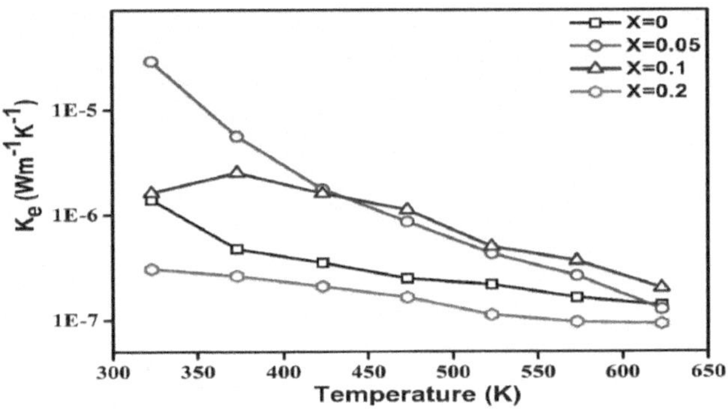

FIGURE 6.4 Temperature-dependent electronic thermal conductivity of $Cu_{2-x}Sb_xTe$ (x = 0, 0.05, 0.1, 0.2) samples [12].

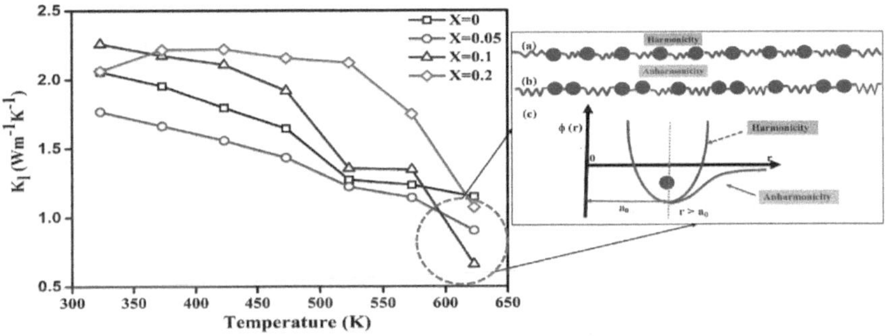

FIGURE 6.5 Temperature dependence of $Cu_{2-x}Sb_xTe$ (x = 0, 0.05, 0.1, 0.2) lattice thermal conductivity [12].

to increased Sb at the Cu location in the lattice. By increasing Sb size, phonon vibration anharmonicity is enhanced, leading to greater phonon–phonon scattering and lowered κ_{lat} without affecting κ_e. Due to its power factor, $Cu_{1.8}Sb_{0.2}Te$ had high ZT values. Sb substitution improves Cu_2Te's thermoelectric characteristics, research shows. Since ZT is identical for SPS and HPHTS processed materials, direct annealing under H_2 improves thermoelectric properties like other procedures.

6.2.2 Alloying/Chemical Doping of Thermoelectric Materials

In low-dimensional materials, quantum confinement changes the band structure. Doping affects carrier concentration, electrical conductivity, and Seebeck coefficient. Mg_2X (X = Si, Sn Ge) compounds have two conduction bands with a small band offset. By modifying the chemical composition of $Mg_2Si_{1-X}Sn_X$ and

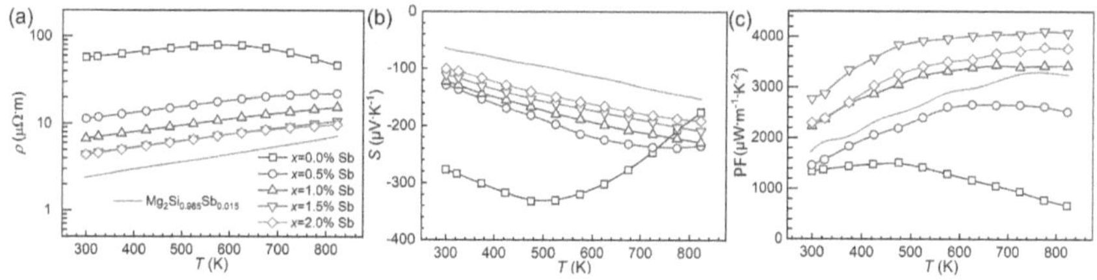

FIGURE 6.6 Electrical properties of $Mg_2(Si_{0.4}Sn_{0.6})_{1-x}Sb_x$ samples: (a) electrical resistivity, (b) Seebeck coefficient, and (c) power factor [14].

$Mg_2Si_{1-X}Ge_X$ solid solutions, the band offset can be altered since Mg_2Sn has inverted light (CL) and heavy (CH) conduction bands. $Mg_2Sn_{0.7}Si_{0.3}$ effectively converges the bands. Band-engineered $Mg_2Sn_{1-x}Ge_x$ compounds had ZT=1.4 at 450°C for bulk $Mg_2Sn_{0.75}Ge_{0.25}$ [13].

Costly, poisonous, or rare bismuth, tellurium, and lead can limit high-performance thermoelectric materials. Good mechanical properties and low density are available in low-cost, eco-friendly alloys like magnesium silicide. N-type materials have been studied for a decade. Thermal performance improves with aliovalent doping, isovalent Si substitution with Sn/Ge, and microstructure manipulation. At the X point, Mg_2Si and Mg_2Sn have opposite light and heavy lowest conduction bands. Isovalent substitution (alloying) in $Mg_2Si_{1-x}Sn_x$ solid solution converges the two bands at x ~ 0.625, increasing mass and power factor. Extreme alloy phonon scattering lowers lattice thermal conductivity. Making $Mg_2(Si,Sn)$ materials at high temperatures is difficult due to Mg's strong reactivity and vapor pressure.

Figure 6.6 shows temperature-dependent electrical transport in $Mg_2(Si_{0.4}Sn_{0.6})_{1-x}Sb_x$ samples, including $Mg_2Si_{0.985}Sb_{0.015}$. Undoped $Mg_2Si_{0.4}Sn_{0.6}$'s electrical resistivity rises and declines at 600 K due to its low carrier concentration, indicating early intrinsic conduction (Figure 6.6a). In $Mg_2(Si_{0.4}Sn_{0.6})_{1-x}Sb_x$ samples, negative Seebeck coefficients indicate primary charge carrier electrons (Figure 6.6b). As with electrical resistivity, the Seebeck coefficient fluctuates with temperature and Sb doping. At high temperatures, undoped $Mg_2Si_{0.4}Sn_{0.6}$'s inherent conductivity lowers the Seebeck coefficient.

Sb doping decreases the bipolar effect and monotonically increases the Seebeck coefficient with temperature. Carrier concentration increases with Sb doping, lowering S. Sb-doped $Mg_2Si_{0.4}Sn_{0.6}$ has a much higher Seebeck coefficient than Mg_2Si, due to solid solution alloying-induced conduction band convergence (Figure 6.4c). In $Mg_2(Si_{0.4}Sn_{0.6})_{1-x}Sb_x$ samples, Si/Sn alloying and Sb doping increase electrical transport. With dominating acoustic phonon scattering, the single parabolic band (SPB) model examines electrical transport characteristics with these equations [14]:

$$S = \frac{-k_B}{e}\left\{\frac{2F_1(\eta)}{F_0(\eta)} - \eta\right\} \tag{6.1}$$

$$n = 4\pi\left(\frac{2m^{*}k_{B}T}{h^{2}}\right)^{\frac{3}{2}} F_{\frac{1}{2}}(\eta) \tag{6.2}$$

where

k_B = Boltzmann constant,

h = Planck constant,

η = Reduced chemical potential, and

$F_j(\eta) = \int_0^{\infty} \frac{x^j}{1+e^{x-\eta}} dx$ is the j-th fermi integral.

6.2.3 Phase Engineering of Thermoelectric Materials

Increased symmetry phase and band structure alterations result in a $1.2\times10^{20}cm^{-3}$ carrier concentration. New thermoelectric performance can be improved by GeSe-based thermoelectric prospects and high-symmetry alloy entropy stabilization. GeSe has a symmetric cubic rocksalt structure at high temperatures but an orthorhombic, highly distorted rocksalt structure at low temperatures. Bond length alterations form van der Waals interlayers. A rhombus resembles stretched opposed cube corners. GeSe alloying with $AgSbSe_2$ distorts the Se sublattice into a rhombohedral shape. Bonding is symmetric cubic over deformed orthorhombic [15].

Cu_2Se pellets SPS-sintered at 850 K had a ZT of 1.82. ZT is enhanced by low κ, resulting from phonon scattering from high-density grain boundaries and dislocations in nanostructure fabrication. SPS-sintering created disc-shaped Cu_2Se pellets for thermoelectric testing. T rises to 850 K, with 0.95–104 S/m^{-1} (Figure 6.7a). As before, a significant fluctuation around 400 K suggests a phase shift occurred during transport property testing with increased T. Most p-type carriers are holes (Figure 6.7b). S rises above bulk Cu_2Se to 296 mV/K^{-1} at 850 K as T increases.

At all temperatures, κ as a function of T (Figure 6.7c) is consistently low (0.4 W/m·K to -0.6 W/m·K, excluding phase transformation values), much lower than bulk Cu_2Se (slightly less than 1 W/m·K). To determine the contributions of κe and κL, we use κe = LσT (L is the Lorenz number) (Figure 6.7d) [16].

6.2.4 Defect Creation of Thermoelectric Materials

TE materials made of InSb perform well at temperatures between 600 K and 800 K, exceeding its Debye temperature (~200–210 K). Due to short-range mass fluctuations and strain field changes, point defects minimize short-wavelength phonon heat at this temperature. The $Ga_xIn_{1-x}Sb$ ($0\leq x\leq0.15$) isoelectronic solution exhibits TE and κL characteristics that scatter short-wavelength phonons while sustaining carrier mobility [17]. Thermal conductivity is determined by lattice phonon (kL) and drift carrier (ke). The lattice conducts electricity and heat separately. The lattice conducts most semiconductor heat. Reduced selective lattice thermal conductivity raises zT. Figure 6.8a–c shows three defect engineering

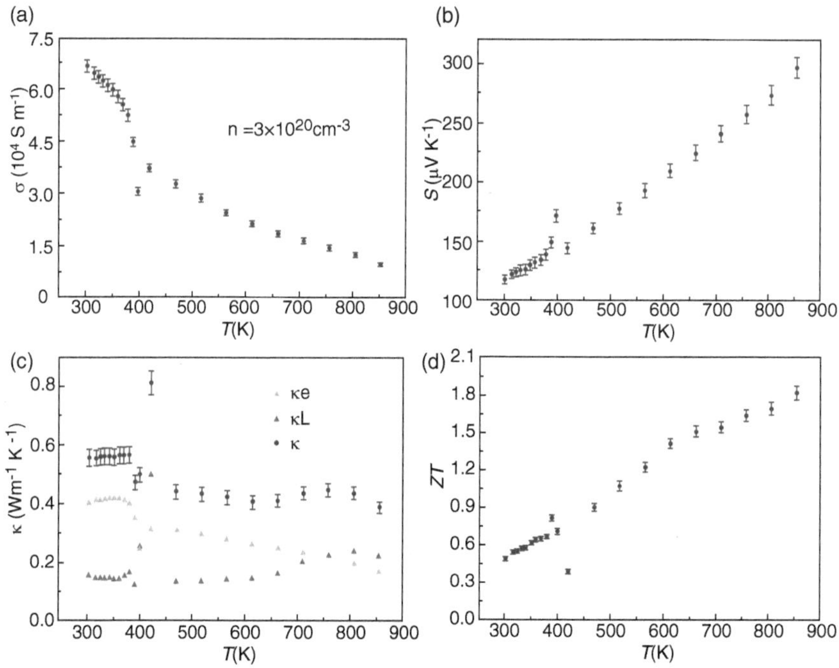

FIGURE 6.7 Typical sintered Cu_2Se pellet thermoelectric characteristics under temperature dependence: (a) electrical conductivity (n=room temperature carrier concentration), (b) Seebeck coefficient, (c) thermal conductivities of Cu_2Se, and (d) calculated ZT values [16].

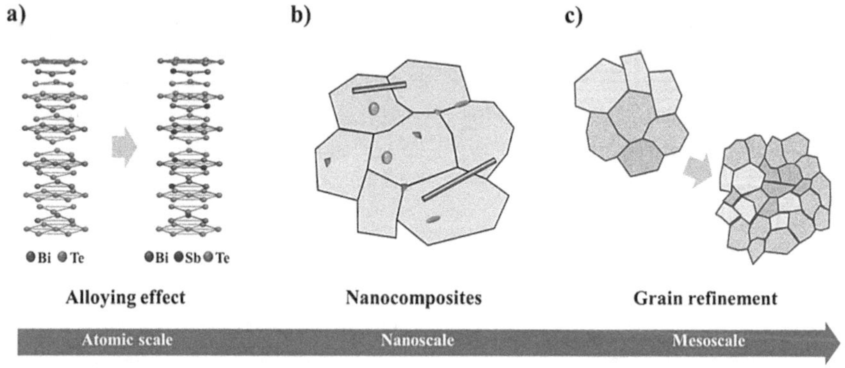

FIGURE 6.8 Diagrams of general thermoelectric property improvement methods: (a) alloying effect, (b) nanocomposites, and (c) grain refinement [18].

methods for lattice thermal conductivity reduction. Atomic-scale alloying uses point impurities (Figure 6.8a). Phonons from point defects scatter at high frequencies and short wavelengths. N- and p-type $Bi_2Te_{3x}Se_x$ and $Bi_{2y}Sb_yTe_3$ are superb room-temperature materials [18].

6.3 MATERIALS COMPATIBILITY AND PROCESS OPTIMIZATION

6.3.1 Materials Compatibility

Compatible thermoelectric materials are used for segmented thermoelectric generators. Calculations for temperature-dependent thermoelectric materials employ thermoelectric potential. When calculating thermoelectric efficiency, compatibility must be considered. Due to thermoelectric material variances, segmented generators must be compatible. If compatibility is [19],

$$s = \frac{\sqrt{1 + zT} - 1}{\alpha T} \tag{6.3}$$

A difference of around a factor of 2 or more means both segments cannot operate efficiently simultaneously, decreasing efficiency. Silicon-based TE is mature, cheap, and flexible. Large-scale integration is possible with 10 nm or smaller TE devices. Low-dimensional materials' thermal characteristics are measured with MEMS [19].

This structure enables micro/nano-scale heat transport and phonon scattering research in TE materials, theoretical zT value optimization recommendations, and faster thermoelectric material and device commercialization. Micro-suspension devices will simplify preparation, improve the measuring principle, error analysis, and measurement accuracy (continuous improvement). For decades, research has improved TE material and device performance. TE was improved by studying synthesis, material components, atomic structures, and device topologies [20]. In Figure 6.9, N-type $Bi_2Te_{0.7}Se_{0.3}$, $Ba_{0.4}In_{0.4}CoSb_{12}$, and P-type Zn_4Sb_3 and Bi_2Te_3 have s values within 2.

6.3.2 Process Optimization

Modern materials research uses microstructure-property connections in synthesis and processing. Traditional experimental research generates a material by trial-and-error from chemical composition and processing conditions and analyzes microstructure and properties. Slow methods based on researchers' expertise and understanding often underestimate materials. Traditional experiment-based methods seem unsuitable for creating novel materials due to the surprising degree of composition and configuration freedom. Significant progress requires chemical insight and serendipity [22].

Although material ZT is important, TEG efficiency is more practical. Processing can affect generator performance with the same thermoelectric materials. The study found a 25% lower Seebeck coefficient in processed generators than in bulk semiconductors. Performance and failure causes of thermoelectric device manufacturing were examined. They used ZT, electrical resistance, and circuit time constant to diagnose thermoelectric devices. A TEG has a thermoelectric module with hot and cold sources and numerous connected units. Electrical plates, n/p semiconductors, and solder layers make thermoelectric units. Loop current is generated by attaching cold and hot sources to the thermoelectric unit's ends (Figure 6.10). For thermoelectric conversion, high ZT semiconductors are best.

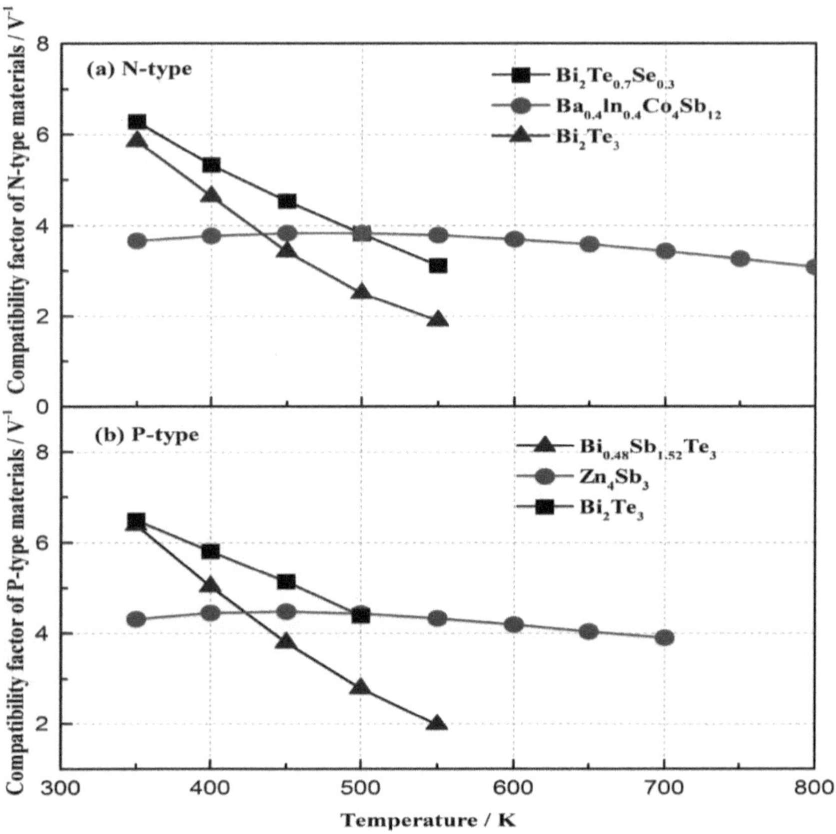

FIGURE 6.9 Materials compatibility factors versus temperature [21].

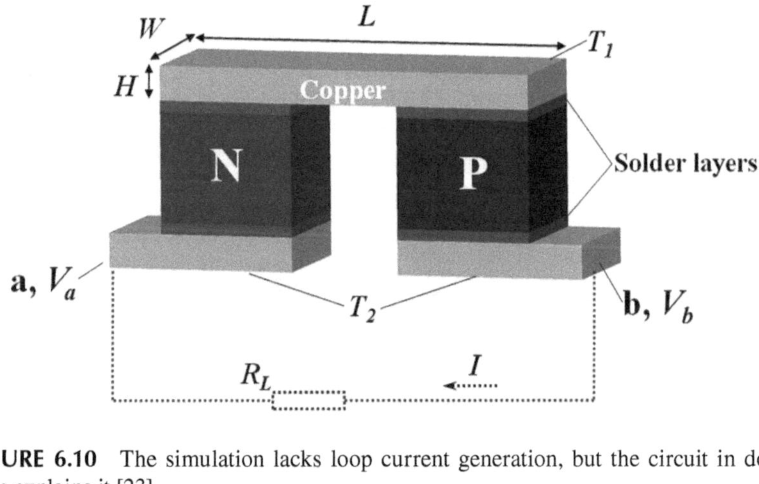

FIGURE 6.10 The simulation lacks loop current generation, but the circuit in dotted lines explains it [23].

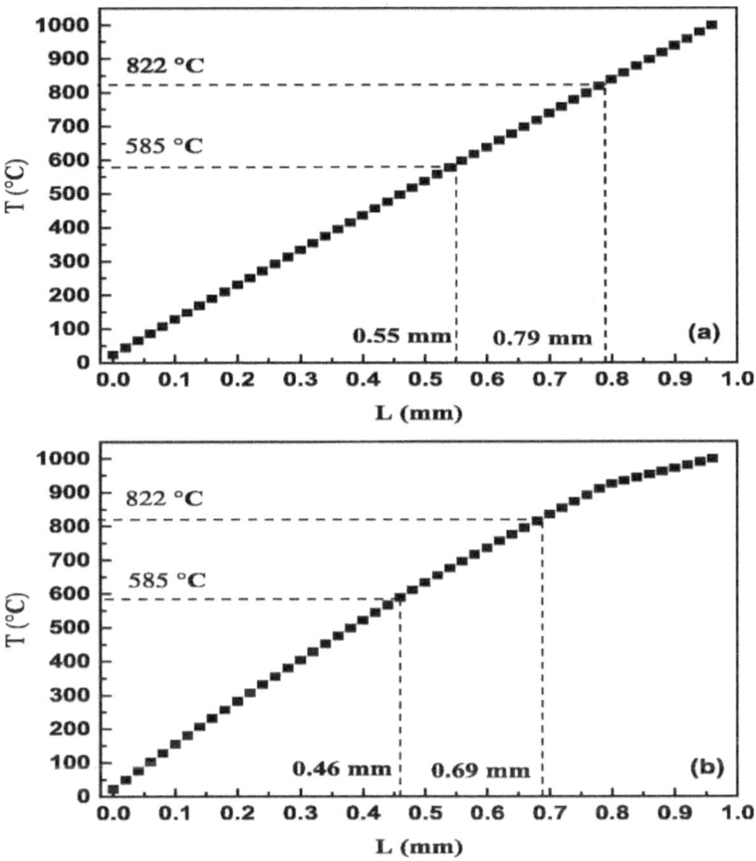

FIGURE 6.11 Temperature distributions along (a) SiGe semiconductor and (b) redesigned coupled semiconductors [23].

Figure 6.11 shows SiGe semiconductor temperature distributions (T_1=1000°C, T_2=20°C). At L distances of 0.55 and 0.79 mm, Bi_2Te_3 melts at 585°C and PbTe at 822°C. Functionally-gradient thermoelectric units are made layer-by-layer from three materials. Figure 6.12 indicates that the new thermoelectric unit has much greater P_{max} and E_{max} than the SiGe unit.

Low-heat conductivity substrates and fillers provide printed TEGs with a higher fill factor than conventional devices. Printer layer thickness, foldable leg length, and application geometry limit printed TEG thickness. The fill factor improves print TEG design and optimization. Optimizations of traditional TEGs commonly use air as filler, which has a low heat conductivity of 0.015 W/m K. This statistic is omitted (rounded to zero) because it is small relative to TE materials' thermal conductivities. For printed TEGs, this approximation fails. Printing thermoelectric devices is adjustable, and roll-to-roll bulk manufacturing cuts costs. Due to the wide range of heat sources, energy harvesting from ambient or waste heat requires distinct device designs for most applications. For energy

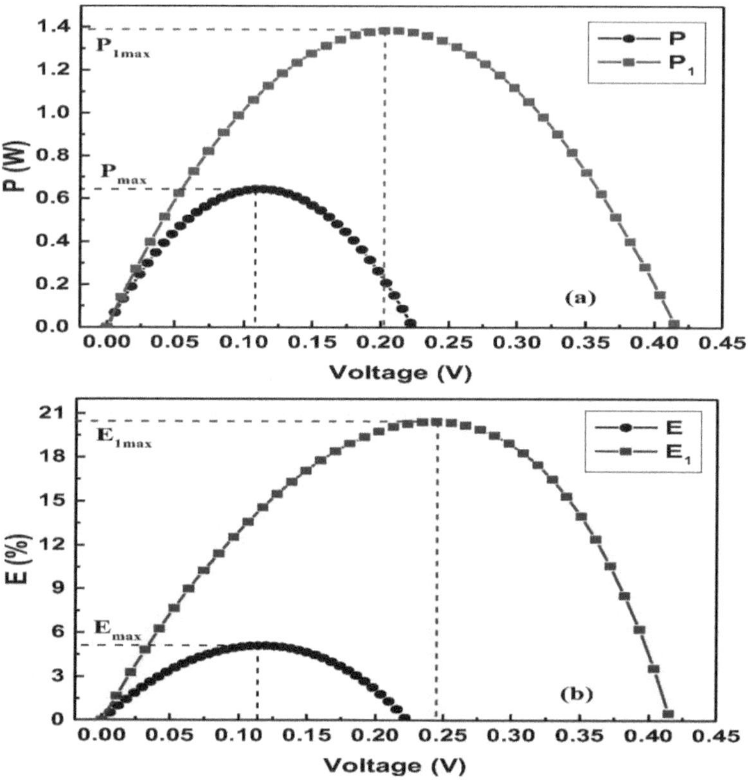

FIGURE 6.12 Comparisons of (a) power output and (b) generating efficiency: P and E for SiGe thermoelectric unit, and P_1 and E_1 for redesigned thermoelectric unit [23].

conversion efficiency, waste heat recovery (WHR) and energy harvesting applications require device design (geometry, substrate material, etc.) and printable TE materials with high zT values.

Figure 6.13 shows κ, dopt, ZT_{max}, and w_{max} for heat exchanger integration with varied κfiller values under fill factor settings. By decreasing the fill factor, κ will converge towards κfiller (Figure 6.13a), resulting in an ideal thickness dopt for small fill factors (Figure 6.13b). A high-κ_{filler} material constricts device thickness adjustment through fill factor. With lower fill factors and larger κ_{filler}, Z_{max}, ZT_{max}, and w_{max} rapidly decrease (Figure 6.13c and d). Thin substrates can give folded devices a higher fill factor than planar devices [24].

6.4 CURRENT STATE AND FUTURE TRENDS IN THERMOELECTRIC EFFICIENCY AND PERFORMANCE

A thermoelectric generator saves heat and boosts efficiency by using an external control system to predict and run thermoelectric modules at full power. This reliable, pollution-free solution helps sustainability, but Carnot's real-time science

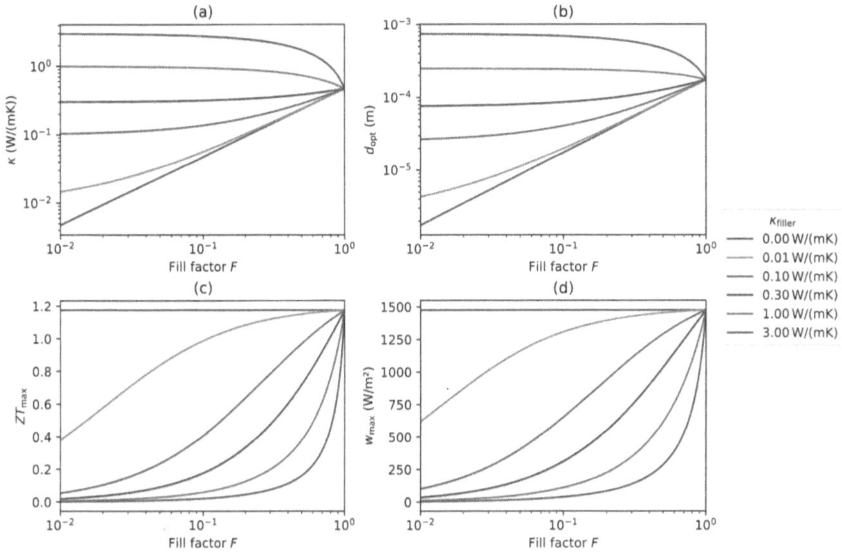

FIGURE 6.13 Heat exchanger impedance-matched device properties: (a) effective thermal conductivity, (b) optimal device thickness, (c) effective ZT value, and (d) maximal output power density for different filler materials as a function of the fill factor [24].

and technology efficiency is superior. This small void slows device functioning by reducing heat-to-electricity conversion. PCM enhances Carnot's thermoelectric conversion. CuS compounds have replaced Bi_2Te_3 due to their low toxicity, high efficiency, and abundance. The best TE material components were found, and powdered components were ball-milled for 10 h. TE material manufacturing continued after the 1-hour heat treatment. TE materials produce less and require many processes to synthesize.

PCMs' high phase change enthalpy at different temperatures makes them helpful for thermal energy storage. PCM type and melting point affect phase change enthalpy. This statistic shares enthalpy release and absorption during two phase changes. PCM needs more phase change enthalpy to improve thermoelectric device efficiency and thermal characteristics. Power is highest from thermoelectrics. PCM's high phase change enthalpy is essential for TE devices. Thicker PCM layers maximize chemical system efficiency. Figure 6.14a and b shows that PCM increases thermoelectric device phase change enthalpy. PCMs increase system phase change enthalpy. Thermal conductivity and phase change enthalpy decreased with low multiwalled carbon nanotube input. With multi-walled carbon nanotube concentration, thermal conductivity rose until it declined linearly [25].

Figure 6.15 shows 1950 thermoelectric materials and recently developed materials. A complex Fermi surface and structure would lower lattice thermal conductivity and increase the power factor in the suggested thermoelectric compound.

Thermoelectric materials could use hierarchical nanocomposites with finer substructures. Current theoretical efforts to manipulate the thermoelectric

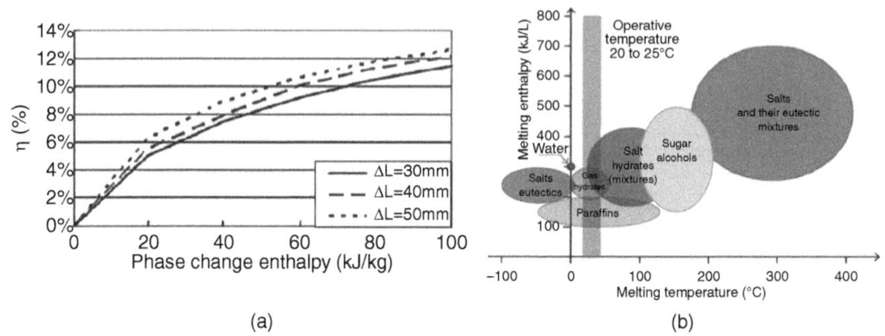

FIGURE 6.14 (a) PCM thickness and phase change enthalpy affect system efficiency. (b) PCM melting enthalpy changes with melting temperature [25].

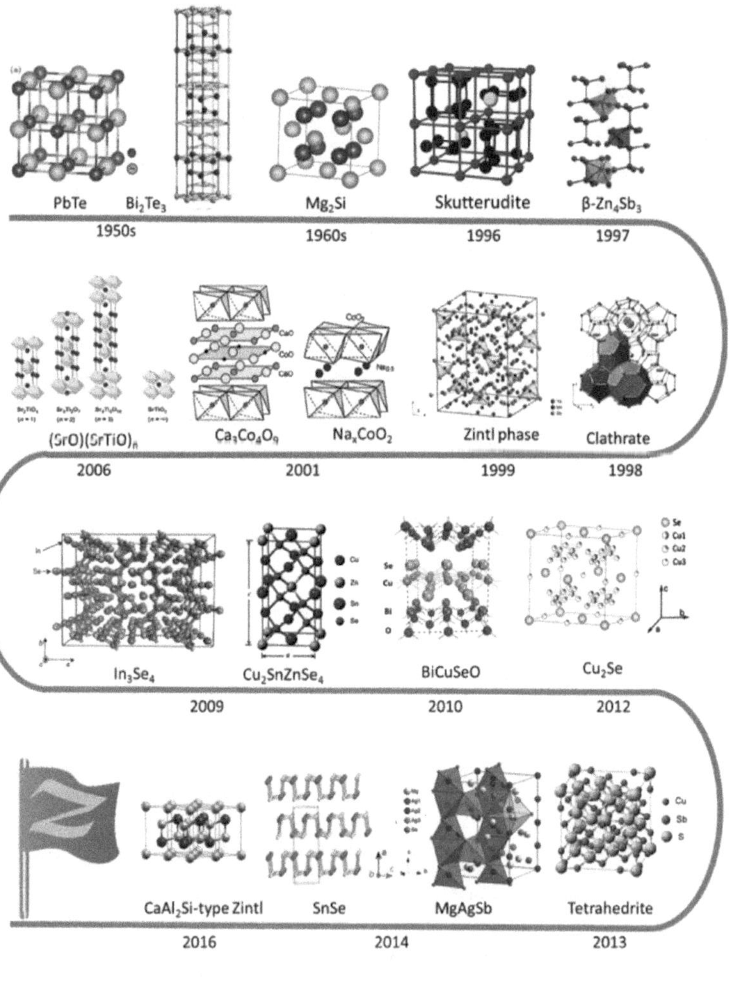

FIGURE 6.15 The timeline of selected thermoelectric materials [26].

figure-of-merit (ZT), in order to enhance its performance, require a high ZT value throughout the entire temperature range, rather than only at its peak. The 3D-printed leg-length gradient-doping optimization seems promising. Pyroelectric and thermogalvanic effects can collect waste heat besides thermoelectric power providers. Flexible thermoelectric materials and technologies are needed for wearable electronics and e-skin after efficiency, efficacy, and dependability [26].

6.4.1 Future Trends

Amorphous materials have the lowest thermal conductivity because the average phonon is the lattice constant (0.2–0.5 nm). To reduce heat conductivity, new physics understanding is needed soon. ZT levels of three or higher seem difficult. The power factor needs to be improved by reducing heat and increasing electron conductivity. To drop thermal conductivity below the amorphous limit, the group velocity or propagating phonon modes must be actively modified. These promising scientific results require theorists and experimentalists to create novel scattering processes and concepts to boost the power factor and lower thermal conductivity. Device assembly, nanostructure, and fabrication determine thermoelectric performance.

Use thermoelectric nanomaterial synthesis technologies for practical thermoelectric applications:

1. Adjustable, high-quality, low-cost thermoelectric capabilities;
2. For machining/device integration, device-controllable nanostructured materials must form dense compacts;
3. The ZT of nanostructured materials should be higher than bulk materials, and eventually; and
4. Compacted nanoscale features should maintain thermal stability. Quantitative research is essential to create and optimize nanostructured materials.

Recent advances in microelectronics have led to smart devices like portable phones, cardiac pacemakers, and exoskeletons that have changed our lifestyles by changing their energy sources. As electronic device energy consumption per unit space reduces, power supply design and implementation become crucial. A wearable TE energy converter converts body heat into electricity to power portable devices. The chemical, physical, and mechanical properties of flexible film- and fiber-based TE energy conversion devices for portable low-power electronic power supplies are still being developed.

1. Wearable gadgets made of film or fiber are too low-output for marketing. Devices that generate more than 5W are needed to power hearing aids, pacemakers, biosensors, etc.
2. Enhance these film- and fiber-based legs for TE energy conversion devices' *inevitable* stretching and compression while humans walk.

3. Creating a commercial prototype using 3D printing or several energy sources. Types excluded for commercial wearable energy collection include:
   i. Advanced 3D printing to create fine micro TE conversion devices for fabric-embedded wearable energy conversion.
   ii. An adequate body-atmosphere temperature difference for tough conditions or space travel is maintained by the wearable TE energy conversion. With flexible and integrated electronics, wearable TE energy converters will present interdisciplinary challenges.

6.5 STRATEGIES FOR ENHANCING DEVICE EFFICIENCY

New TE materials and performance optimization have been achieved using many material synthesis processes (Figure 6.16a). There are chemical and physical synthesis methods. Based on synthesis, TE materials are bulk, thin films, nanowires, etc. Figure 6.16b shows series-connected P–N material TE device manufacture. Splicing and welding rigidify bulk TE devices. New printing methods can tailor any device and save materials. Micro-nano manufacturing integrates high-density TE modules for mass production and high voltage [27].

Modern thermoelectric materials are mostly inorganic Bi_2Te_3, which is expensive and difficult to manufacture.

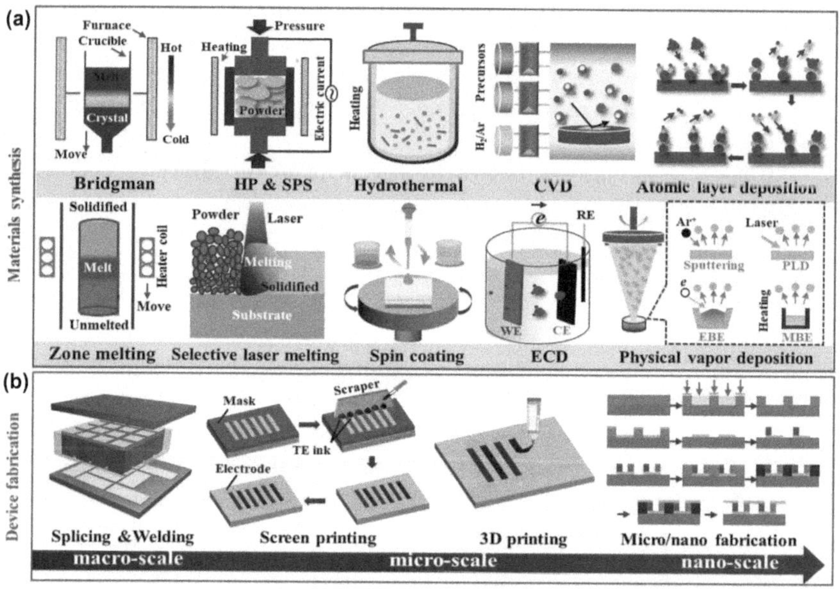

FIGURE 6.16 (a) Hot pressing, spark plasma sintering, and CVD are abbreviations. Electrodeposition, work, reference, and counter electrodes are ECD, RE, WE, and CE. Pulsed laser deposition, electron beam evaporation, and molecular beam epitaxy are used. (b) Device fabrication methods [27].

1. Mg_3Bi_2, Ag_2S, and PEDOT:PSS molecules are prevalent. Low material performance limits their application. Create more affordable, easy-to-prepare, high-performance thermoelectric materials.
2. TEC device manufacture is complex and time-consuming, utilizing classic methods such as sintering, cutting, and welding for bulk devices and sputtering and deposition for thin-film. Rising demand requires innovative printing and magnetron sputtering technologies.
3. Future device design focuses on miniaturizing flexible TECs for superior cooling performance and size reduction. Materials and device commercialization is problematic despite TECs' superior cooling capability. While traditional Bi_2Te_3-based thermoelectric materials are costly, new low-cost near-room temperature materials are urgently needed. Traditional device manufacturing needs delicate, time-consuming cutting and soldering.

PbSe compounds, both n- and p-type, enhance thermoelectric performance through meticulous carrier concentration, electronic band structure engineering, and phonon engineering. It will take time to commercialize PbSe.

6.6 FUTURE OUTLOOK

i. Optimize carrier concentration at all temperatures. Temperature determines the ideal carrier concentration for p-type PbSe, with a power law of $\sim T^3/2$. Due to complete ionization below room temperature, traditional dopants have invariant carrier concentrations at high temperatures. Manipulating shallow and deep impurity levels or PbSe electrical band structures can fix this problem.
ii. Enhancing PbSe service stability or low peak ZT temperatures. When Se evaporates, PbSe loses weight at approximately 750 K.

The most common thermoelectric elements are bulk alloys like Bi_2Te_3, PbTe, SiGe, and $CoSb_3$, with a ZT around 1 and attractive prices. A TEG compared standard Bi_2Te_3-based and advanced quantum well materials for car waste heat recovery. Powerful quantum-well generators outperform Bi_2Te_3 generators. Thermoelectric generator fuel economy for light-duty passenger cars. A FEM simulation demonstrates that nanostructured bulk half-Heusler thermoelectric modules can increase fuel efficiency by 2.5%. In addition to TE materials, other measures are needed to increase TE performance. Copper electrodes link p- and n-type thermoelectric devices in series and thermally in parallel between two electrically insulated ceramic panels in the thermoelectric module.

High-efficiency topological connections between TEMs can remove this constraint. ATEG system TEM electric topology genotype (GA) optimization. Downward TEM output is possible. An Mg_2Si/Zn_4Sb_3 thermoelectric generator is employed for high-temperature zones and Bi_2Te_3 for low-temperature zones since

TE materials have variable optimal operating temperatures. At 330°C exhaust temperature and 0.22 kg/s mass flow rate, pure Mg_2Si/Zn_4Sb_3 TEG generates 1050 W and half-portion 1200 W. Three heat exchangers were examined to determine how winglet vortex plate configuration influences exhaust-based gas-to-liquid TEG performance.

Two methods have been used to raise zT in TE materials:

1. Reducing κL near the amorphous limit through defect engineering or nanostructure does not impact electrical performance. Optimization of n or band engineering can improve power factor (PF, $S^2\sigma$) while also improving S and σ.
2. The absence of κe in $S^2\sigma$ limits its ability to completely quantify the impact of electron transport on zT and tune at the same level as zT.

Thermoelectric technology reverses thermal-to-electricity conversion. High zT materials at low temperatures can recycle significant amounts of low-grade heat energy (below 200°C) that ordinary heat engines cannot handle. Materials should be content-rich, non-toxic, and low in synthesis.

To increase conversion efficiency, device construction must optimize module size and physical interface. The device manufacturing technology must save raw materials, mass produce, and be cost-effective. Design adaptive thermal management systems improve electronic device performance and longevity. Power generation often requires TE device heat sources. Waste heat from equipment, water, gas, etc. can be recycled to save energy. Another possibility is using natural heat sources like solar, land, ocean, and radioisotopes.

TE devices can receive biological heat (heat changes in live activities, especially from the human body), chemical heat (endothermic and exothermic), or heat converted from other signals to power themselves. With radiative cooling, TE devices can generate power all day. Soon, TEG will generate power continuously using a new photothermal and radiative cooling technology. The inverse effect of electricity generation allows TE devices to heat and cool locally. Additionally, the TE cooler and solar energy system can desalinate without requiring more energy [27].

Material interconnections impact transport characteristics (σ, S, and k). σ and ke share a positive correlation, while σ and S have a negative correlation. K_l is independent but positively correlated with σ due to structural symmetry. These correlations complicate ZT optimization. As charge concentration (n) increases, σ and k improve, while S drops. Optimizing n ($10^{19}cm^{-3}$) by defect engineering or doping enhances ZT value. However, dopants reduce charge carrier mobility, interrupting PF optimization.

A challenging relationship between these parameters requires material attributes to be balanced in order to maximize ZT. State-of-the-art TE materials' ZT values since 2000 are presented in Figure 6.17.

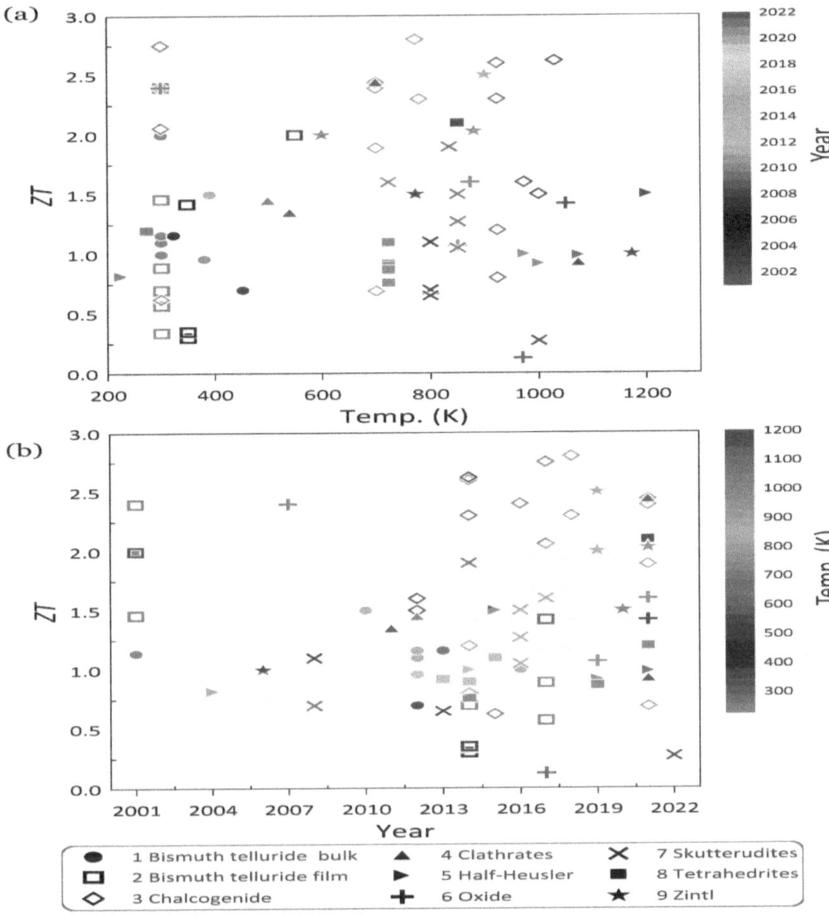

FIGURE 6.17 The merit of modern TE materials since 2000: (a) different inorganic TE materials' ZT values and (b) inorganic TE material timelines [28].

REFERENCES

1. Yang, L., Chen, Z. G., Dargusch, M. S., & Zou, J. (2018). High performance thermoelectric materials: progress and their applications. *Advanced Energy Materials*, *8*(6), 1701797.
2. Shi, W., Wang, D., & Shuai, Z. (2019). High-performance organic thermoelectric materials: theoretical insights and computational design. *Advanced Electronic Materials*, *5*(11), 1800882.
3. Ryu, B., Chung, J., Kumagai, M., Mato, T., Ando, Y., Gunji, S., & Park, S. (2023). Best thermoelectric efficiency of ever-explored materials. *Iscience*, *26*(4), 106494.
4. Zhu, H., & Xiao, C. (2018). Strategies for discovery and optimization of thermoelectric materials: role of real objects and local fields. *Frontiers of Physics*, *13*, 1–8.

5. Shi, W., Wang, D., & Shuai, Z. (2019). High-performance organic thermoelectric materials: theoretical insights and computational design. *Advanced Electronic Materials*, *5*(11), 1800882.
6. Kandemir, A., Ozden, A., Cagin, T., & Sevik, C. (2017). Thermal conductivity engineering of bulk and one-dimensional Si-Ge nanoarchitectures. *Science and Technology of Advanced Materials*, *18*(1), 187–196.
7. Shtern, M. Y. (2019). Current trends in improving the efficiency of thermoelectric generators. In *2019 IEEE Conference of Russian Young Researchers in Electrical and Electronic Engineering (EIConRus)*, Saint Petersburg and Moscow, Russia (pp. 1914–1919). IEEE.
8. Wan, S., Song, Q., Chen, H., Zhang, Q., Liao, J., Xia, X., ... & Chen, L. (2023). High-efficiency segmented thermoelectric power generation modules constructed from all skutterudites. *Cell Reports Physical Science, 4*, 101651.
9. Jha, P. K. (2021). Strategies for improving efficiency of thermoelectric materials. In Kumar R. & Singh R. (eds.). *Thermoelectricity and Advanced Thermoelectric Materials* (pp. 117–138). Woodhead Publishing.
10. Shtern, M., Sherchenkov, A., Shtern, Y., Borgardt, N., Rogachev, M., Yakubov, A., ... & Murashko, D. (2023). Mechanical properties and thermal stability of nanostructured thermoelectric materials on the basis of PbTe and GeTe. *Journal of Alloys and Compounds*, *946*, 169364.
11. Mamur, H., Üstüner, M. A., Korucu, H., & Bhuiyan, M. R. A. (2023). A review of the performance evaluation of thermoelectric nanostructure materials $Bi_{2-x}Sb_xTe_3$ ($0.20 \leq X \leq 1.80$). *Cleaner Chemical Engineering*, *6*, 100101.
12. Rajkumar, R., Mani, J., Nedunchezhian, A. A., Sidharth, D., Radha, S., Arivanandhan, M., & Anbalagan, G. (2023). Electrical and thermal transport properties of Sb substituted Cu_2Te nanostructures for thermoelectric applications. *Inorganic Chemistry Communications*, *151*, 110622.
13. Safavi, M. (2021). *Investigation of thermoelectric properties of magnesium-based metal alloys coatings deposited by sputtering process*. Doctoral dissertation, Université Bourgogne Franche-Comté.
14. Wang, B., Zhao, H., Li, J., Zhang, B., Wang, D., Chen, C., ... & Tian, Y. (2023). Enhanced thermoelectric performance of Sb-doped $Mg_2Si_{0.4}Sn_{0.6}$ via doping, alloying and nanoprecipitation. *Journal of Materiomics, 10*, 285–292.
15. Huang, Z., Miller, S. A., Ge, B., Yan, M., Anand, S., Wu, T., ... & Bao, X. (2017). High thermoelectric performance of new rhombohedral phase of GeSe stabilized through alloying with $AgSbSe_2$. *Angewandte Chemie International Edition*, *56*(45), 14113–14118.
16. Yang, L., Chen, Z. G., Han, G., Hong, M., Zou, Y., & Zou, J. (2015). High-performance thermoelectric Cu_2Se nanoplates through nanostructure engineering. *Nano Energy*, *16*, 367–374.
17. Du, Z., He, J., Chen, X., Yan, M., Zhu, J., & Liu, Y. (2019). Point defect engineering in thermoelectric study of InSb. *Intermetallics*, *112*, 106528.
18. Park, J. G., & Lee, Y. H. (2016). High thermoelectric performance of Bi-Te alloy: defect engineering strategy. *Current Applied Physics*, *16*(9), 1202–1215.
19. Snyder, G. J. (2003). Design and optimization of compatible, segmented thermoelectric generators. In *Proceedings ICT'03. 22nd International Conference on Thermoelectrics (IEEE Cat. No. 03TH8726)*, La Grande Motte, France (pp. 443–446). IEEE.
20. Ma, Z., Wei, J., Song, P., Zhang, M., Yang, L., Ma, J., ... & Wang, X. (2021). Review of experimental approaches for improving zT of thermoelectric materials. *Materials Science in Semiconductor Processing*, *121*, 105303.

21. Zhu, L., Li, H., Chen, S., Tian, X., Kang, X., Jiang, X., & Qiu, S. (2020). Optimization analysis of a segmented thermoelectric generator based on genetic algorithm. *Renewable Energy*, *156*, 710–718.
22. Wang, Z. L., Adachi, Y., & Chen, Z. C. (2020). Processing optimization and property predictions of hot-extruded Bi–Te–Se thermoelectric materials via machine learning. *Advanced Theory and Simulations*, *3*(1), 1900197.
23. Ma, Q., Fang, H., & Zhang, M. (2017). Theoretical analysis and design optimization of thermoelectric generator. *Applied Thermal Engineering*, *127*, 758–764.
24. Rösch, A. G., Franke, L., Mallick, M. M., & Lemmer, U. (2023). Optimizing printed thermoelectric generators with geometry and processibility limitations. *Energy Conversion and Management*, *279*, 116776.
25. Pandey, R., Thapa, P., Kumar, V., Zhu, Y., Wang, N., Bystrzejewski, M., & Tiwari, S. K. (2022). Updates in phase change materials for thermoelectric devices: status and challenges. *Materialia*, *21*, 101357.
26. Liu, W., Hu, J., Zhang, S., Deng, M., Han, C. G., & Liu, Y. (2017). New trends, strategies and opportunities in thermoelectric materials: a perspective. *Materials Today Physics*, *1*, 50–60.
27. Wu, Z., Zhang, S., Liu, Z., Mu, E., & Hu, Z. (2022). Thermoelectric converter: strategies from materials to device application. *Nano Energy*, *91*, 106692.
28. Giri, K., Wang, Y. L., Chen, T. H., & Chen, C. H. (2022). Challenges and strategies to optimize the figure of merit: keeping eyes on thermoelectric metamaterials. *Materials Science in Semiconductor Processing*, *150*, 106944.

7 Case Studies
Examples of Thermoelectric Materials and Devices, Along with Real-World Applications and Research Success Stories

7.1 INTRODUCTION

Wearable and adaptable thermoelectric energy harvesting technologies are becoming popular. These materials efficiently and sustainably generate power from waste, body, or ambient heat. Lightweight organic thermoelectric compounds, compatible with flexible substrates, can increase energy sector breakthroughs. Many new thermoelectric materials, which generate electricity from waste heat, are drawing interest. Temperature gradients create voltage. The conversion efficiency of these alternate thermoelectric materials depends on their dimensionless figure of merit (ZT).

ZT is defined as Equation (7.1):

$$ZT = \left[\frac{\alpha\sigma}{\kappa}\right]T \tag{7.1}$$

where

α = Electrical resistivity,

κ = Thermal conductivity, and

T = Absolute temperature.

High electrical conductivity (low Joule heating), a large Seebeck coefficient (large potential difference), and low thermal conductivity are essential for thermoelectric material performance. The ZT measures TE device efficiency [1].

Sustainable development research has focused on thermoelectric materials' direct waste heat-to-electricity conversion. Organic thermoelectric (OTE) materials are ideal for wearable and portable electronics due to their low density, flexibility, and eco-friendliness. A global sustainable energy solution may include thermoelectric materials, which can generate power from waste heat or be solid-state Peltier coolers. It is hard to identify materials with improved thermoelectric efficiency due to competing qualities.

DOI: 10.1201/9781003639343-7

Complex thermoelectric materials, especially nanoscale ones, are entering a new era due to enhanced synthesis and characterization methods. Improvements to thermopower and thermal conductivity are reviewed. Waste heat recovery, portable electricity, cooling, and refrigeration use organic thermoelectric materials. Increased thermoelectric efficiency and stability under diverse operating conditions are needed.

Energy harvesting is possible with higher ZT values because thermoelectric conversion is more efficient. These materials significantly reduce energy use and greenhouse gas emissions [2–4].

Common organic compounds include conductive polymers, organic-inorganic hybrids, and small organic molecules. Electrical conductivity and thermopower increase with thermoelectric materials. They are flexible and lightweight, have low heat conductivity for thermoelectric efficiency, are scalable for cost-effective printing and coating, and are biocompatible [5].

Recent research on alternative thermoelectric materials, their integration into numerous applications, and their potential to improve energy efficiency and sustainability are covered in this chapter. It also suggests future research and challenges.

7.2 ENERGY HARVESTING USING ALTERNATIVE THERMOELECTRIC MATERIALS

Organic substances, skutterudites, silicon-germanium alloys, lead telluride, and bismuth telluride show potential for energy conversion efficiency. Each of these materials improves thermoelectric devices with distinct properties. TE heat engines are shown in Figure 7.1. In the Peltier effect, TE junctions are heated or cooled by current flow. When heat crosses the junction, the Seebeck effect generates electricity. Practical TE generators connect multiple junctions in series for better voltage and heat distribution.

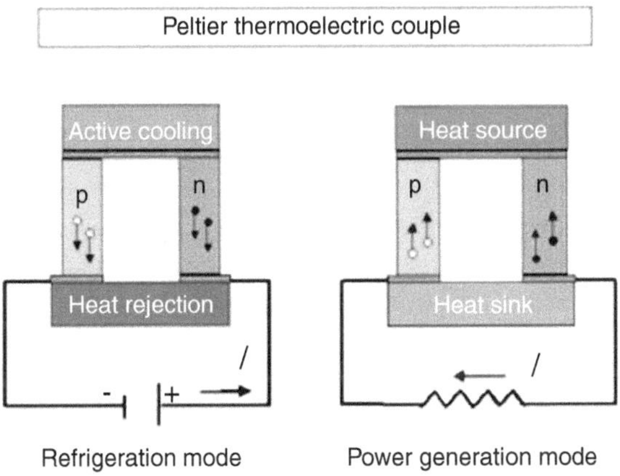

FIGURE 7.1 Schematic of an n-type (electrons, closed circle arrows) and p-type (holes, open circle arrows) thermoelectric couple or module. Power generation (Seebeck effect) and electronic cooling (Peltier effect) are shown in this diagram of the thermoelectric module [6].

Small molecules and carbon nanoparticles improve thermoelectric properties. ZT is high for n-doped small molecules. Theories propose that toxic-element-free heavy metal chalcogenides have a ZT, making them thermoelectric. Graphene and carbon nanotubes' high electrical conductivity benefits thermoelectric applications.

Energy harvesting using organic thermoelectric materials (OTEMs), especially for heat conversion, has garnered attention. This technology suits wearable electronics and low-power sensors because it is adaptable, portable, and expandable [7]. The low thermal conductivity of OTEMs makes them ideal for energy harvesting [8]. Recent advances in organic thermoelectric materials make them attractive for energy harvesting and waste heat recovery [9]. These materials can power wearables and capture vehicle energy [10].

Their non-toxicity and safe materials make them eco-friendly. Organic materials are cost-effective and suitable for various applications [11]. Energy conversion with OTEMs is promising and affordable, being lightweight, adaptable, and configurable. Green, readily available materials endure longer than conventional ones. Price and processing ease make them flexible and mass-producible. These materials are biocompatible, safe, and medically beneficial. With high efficiency, they generate power from minor temperature differences [3].

OTEs generate power from material temperature variations using the Seebeck effect. The materials' Seebeck coefficient, electrical conductivity, and thermal conductivity affect conversion efficiency [12]. Flexible mechanical properties and solution processing make organic materials suitable for large-scale applications. Waste heat-to-power OTEGs are cheaper than thermoelectric generators and are suitable for industrial use [13].

Due to its electrical structure, poly(3,4-ethylenedioxythiophene): poly (styrenesulfonate) (PEDOT:PSS) is a conducting polymer with a high Seebeck coefficient. PEDOT:PSS's extreme heat conductivity limits thermoelectric performance [14,15]. Another band-structured conducting polymer, poly(3-hexylthiophene) (P3HT), has a high Seebeck coefficient. Low thermal conductivity makes P3HT a plausible thermoelectric material. Pricey and difficult to produce, P3HT is limited in thermoelectric devices. P3HT is sensitive to radiation and high temperatures, making it unsuitable for several applications.

Poly(3,4-ethylenedioxythiophene) is another promising thermoelectric conductor. PEDOT is promising for thermoelectric energy conversion due to its high electrical and low thermal conductivity. Stability and mechanical flexibility make it ideal for thermoelectric materials. PEDOT is non-toxic and safe for various uses. Affordable pricing improves its commercial appeal [15].

PEDOT:PSS, a conductive polymer and tellurium nanowire hybrid, can increase its thermoelectric properties by modifying the nanoparticle shape and adding polar solvents. The power factor improves when PEDOT:PSS doped with polar solvents transitions from hopping transport to carrier scattering. This increases electrical conductivity without compromising thermopower. Optimizing settings can boost the power factor of the novel thermoelectric material to beyond 100 $\mu W\ m^{-1}\ K^{-2}$, surpassing pure PEDOT:PSS [5].

Nanostructuring, doping, and molecular design improve organic thermoelectric performance. Optimizing molecular and interfacial designs enhances ZT in organic-inorganic composites. Proof-of-concept studies propose using organic and inorganic materials in flexible and micro-thermoelectric generators.

Effectiveness and efficiency require greater research on these materials. Spreading these technologies requires cost-effective fabrication methods [4,8].

These fascinating electrical transport properties have garnered attention for thermoelectric materials, especially skutterudite compounds. Skutterudites' cage-like crystals limit thermal conductivity and increase electrical conductivity. Researchers are doping skutterudites with various materials to improve these features. Many studies underline the need to understand the effects of skutterudite framework doping for thermoelectric optimization.

Innovative waste heat recovery systems and automotive and industrial cooling solutions may result from these technologies [16–18]. Sn was added to $Yb_{0.3}Co_4Sb_{12}$ alloy to change electrical transport behavior and improve thermoelectric properties by Qin et al. [16]. Sn modification increases Seebeck coefficients without modifying power factors by reducing carrier concentration and ionized impurity dispersion. Adding Sn lowers electrical and lattice thermal conductivity. It was found that altering electronic heat conductivity enhances thermoelectric properties [16].

Qin et al. evaluated Si-doped Yb-filled $CoSb_3$-based skutterudites' thermoelectric performance. Si nanoparticles in $Yb_{0.3}Co_4Sb_{12}$-based skutterudites increased power factor and decreased electronic heat conductivity. CoSi nanoparticles inhibiting phonon transport raised the ZT value of the $Yb_{0.3}Co_4Sb_{12}/_{0.1}Si$ sample to 1.37 at 823 K. Though residual $YbSb_2$ phases presented issues, Si doping raised $ZT_{average}$ by 21% for $Yb_{0.3}Co_4Sb_{12}/_{0.1}Si$ [17].

Si-doped Yb-filled $CoSb_3$-based skutterudites were studied for thermoelectric performance by Qin et al. [17]. Si doping in $Yb_{0.3}Co_4Sb_{12}$-based skutterudites produces CoSi nanoparticles, increasing power factor and decreasing ETC. Due to CoSi nanoparticles impeding phonon transmission, the $Yb_{0.3}Co_4Sb_{12}/_{0.1}Si$ sample's ZT increased to 1.37 at 823 K. The $Yb_{0.3}Co_4Sb_{12}/_{0.1}Si$ sample had 21% higher $ZT_{average}$ after Si doping, despite residual $YbSb_2$ phases [17].

For high-temperature thermoelectric applications, SiGe, PbTe, Bi_2Te_3, and related alloys have been extensively studied to increase energy conversion efficiency. These alloys' thermal and electrical qualities make them valuable for power generation in harsh environments like space missions [19]. Research reveals that nano-structuring, nanoengineering, and alloy composition adjustment improve performance. Electronic cooling, portable cooling, and small-scale power generation are feasible [20–22].

Alternative thermoelectric materials are intriguing for innovation because they are more flexible, mechanically sound, and thermoelectrically efficient than inorganic materials. Organic semiconductors, hybrid composites, tiny molecules, and carbon nanomaterials enable new energy harvesting and conversion applications for more sustainable thermal energy use.

7.3 HYBRID APPLICATIONS AND SYSTEM INTEGRATION OF THERMOELECTRIC MATERIALS

Studying thermoelectric materials' performance can indicate their role in hybrid systems. Thermoelectric hybrids reduce carbon emissions and increase energy efficiency. New waste heat recovery, cooling, and power-generating applications are examined, and material engineering and system design are suggested to improve performance.

For efficient thermal-to-electric conversion, solar thermoelectric-photovoltaic hybrids and vehicle waste heat harvesting require multiscale thermoelectric materials with non-equilibrium synthesis. Photovoltaic and thermoelectric energy conversion helps these systems [23]. Energy and carbon emissions can be decreased using hybrid systems. Multi-scale thermoelectric materials can utilize solar and wind energy [24]. Organic and hybrid thermoelectrics generate renewable electricity [11]. Hybrid systems with thermoelectric materials can enhance energy conversion and enable new uses. Bismuth, silicon-germanium, and lead tellurides work well, whereas organic and hybrid materials are flexible and suitable for low temperatures. Mobile platform cooling and solar thermoelectric-photovoltaic systems show their viability.

Nanocomposites can increase thermoelectric performance. Combining MoS_2 with MoS_2 nanoribbons improves thermoelectric properties. Electricity and Seebeck coefficient increase when the two nanoribbons work together. These nanocomposites are more stable and thermally and electrically robust. Improved mechanical properties and flexibility make them suitable for numerous applications [25].

Carbon allotrope hybrids optimize material properties and reduce heat conductivity for thermoelectric applications. These hybrids are lightweight and affordable, making them suitable for various applications. Their practicality is enhanced by their renewable material composition [26].

Thermoelectric hybrids save energy and improve sustainability. Using thermoelectric devices to convert waste heat into power minimizes carbon emissions and increases energy efficiency. These materials generate electricity from solar and wind power, reducing fossil fuel use and climate impact. Fuel efficiency and pollution reduction are improved by thermoelectric materials in car systems. Industries may become more sustainable with thermoelectric technology. In enterprises and processes, thermoelectric materials can generate power from waste heat, reducing carbon emissions [27].

Improving thermoelectric material stability, durability, scalability, and cost-effectiveness in hybrid systems is essential. Researchers are exploring new encapsulating and coating technologies to stabilize and protect these materials. They are also developing stronger, more efficient materials. Energy harvesting, automotive, and aerospace applications are considering thermoelectric materials. Future research on thermoelectric materials will focus on more efficient materials, production processes, and device design for specialized applications, enabling their use in a broader range of applications [28].

7.4 NANOSCALE AND ADVANCED THERMOELECTRIC DEVICES

Nanoscale and advanced thermoelectric devices are necessary for energy efficiency and sustainability. These devices use waste heat to generate energy, utilizing thermally and electrically optimized materials. Recent research has examined new ways to improve these traits. Nanocomposites with optimized interfaces lower heat conductivity and increase the power factor [29]. Nanoscale defect patterns in n-type PbSe affect charge and phonon transmission, making it thermoelectric. Altering these flaws improves materials' electrical and thermoelectric performance. Additionally, nanoscale defects can improve energy storage materials [30].

Ibáñez et al. created thermoelectric nanocomposites using a bottom-up strategy, resulting in improved thermal conductivity, charge carrier concentration, and efficiencies up to 1.7 at 850 K [31]. To improve thermoelectric materials, superlattices, quantum dots, nanowires, and nanocomposites decouple electrical, thermal, and Seebeck coefficients [32]. Thermal and electrical conductivities are optimized by decoupling, enhancing efficiency. Additionally, nanoparticles can increase thermoelectric materials' operational temperature range. Efficiency helps energy, cooling, and refrigeration companies. Solar panels, transistors, batteries, and electronics benefit from nanomaterials. High surface area and pore volume make them ideal catalysts and adsorbents.

Self-powered wearable sensors and medical monitoring systems benefit from organic thermoelectric materials [8]. These devices generate power and improve usability and performance using body heat. Organic thermoelectric materials' low thermal conductivity reduces heat loss and improves energy conversion. Organic thermoelectric materials can be tuned for application optimization. Altering their chemical structure and content can improve thermoelectric performance for wearable devices and self-powered systems [33]. Recently developed hybrid thermoelectric materials and systems promise energy-efficient applications.

7.5 SUCCESS STORIES IN THERMOELECTRIC RESEARCH

Thermoelectric solid-state energy converters utilize their thermal, semiconducting, and electrical properties to convert waste heat into power, or for cooling and heating. They can compete with air-conditioning compressors and heat pumps or be used in automobile seats, night-vision systems, and electrical enclosure cooling. Improving energy-conversion efficiency and system architecture can help utilize more thermoelectric materials. Demonstrations and prospective uses of thermoelectric technology highlight these concepts [6]. Thermoelectric research promotes dimensionless material merit. Due to the demand for efficient energy conversion methods, thermoelectric research has advanced.

Advances include nano structuring [1], new materials [34], and improved synthesis. Some innovative methods and high-performance materials and applications have advanced thermoelectric research [35]. Hybrid thermoelectric systems, such as solar thermoelectric-photovoltaic and vehicle waste collection systems, boost energy conversion. Thermoelectric modules for energy scavenging and supplementary cooling have enhanced mobile platform system efficiency [36].

For organic thermoelectric materials, TEGs are crucial. Portable electronics and sensors use body-heat generators. Flexible, ultra-thin films and fibers that integrate into clothes and accessories are new developments. According to case studies, wearable TEGs can power heart rate monitors and generate electricity from body heat during activity [8].

The PbS–Ag system study discovered that Ag nanodomains inhibit phonon propagation and supply electrons to the PbS semiconductor, lowering charge transfer inter-grain energy barriers. Therefore, PbS–Ag nanocomposites have lower thermal conductivities and higher charge carrier concentrations and mobilities than PbS nanoparticles. Improvements in material transport result in a 1.7 thermoelectric figure of merit at 850 K [31].

Building materials using organic thermoelectric elements conserve energy in homes and businesses. Thermoelectric paints and roof coatings capture solar heat, and HVAC systems use thermoelectric panels to conserve energy. Case studies show energy retrofits and smart building designs using thermoelectric materials to manage energy.

Hybrid and electric vehicle batteries and fuel efficiency can be improved by organic thermoelectric materials processing exhaust heat. Energy recovery technologies improve fuel economy in case studies.

Future concerns include the sustainable, stable, and efficient integration of organic thermoelectric materials. New material compositions, advanced characterization methods, and research collaboration may accelerate innovation and commercialization in this field. Organic thermoelectric materials have enormous potential for sustainable energy harvesting in many industries, but further research is needed.

7.5.1 Success Stories in Thermoelectric Research

Global energy needs and environmental concerns have boosted interest in thermoelectric conversion technology. High-performance thermoelectric materials have improved. A thermoelectric generator generates electricity from waste heat. Peltier coolers and thermoelectric generators have no moving parts, unlike heat engines, freezers, and air conditioners. These devices could endure 25–30 years without repair.

Even though less efficient than generators, thermoelectric devices can capture energy in small applications. These devices capture automotive, factory, and human waste heat for Seiko and Citizen wristwatches. High-efficiency thermoelectric devices are being created using nanowire, thin film, and bulk nanostructured materials.

The orthorhombic phase is stable for SnS, SnSe, GeS, GeSe, SnTe, PbS, PbSe, PbTe, SnS2, and SnSe2. These compounds are narrow band gap semiconductors in the 0.1–2.8 eV energy range, according to electronic band gap calculations. They exhibit direct cubic band gaps and indirect orthorhombic band gaps for tin, lead, and germanium chalcogenides. The Seebeck coefficient, power factor, electrical, and thermal conductivity of these substances are estimated using Boltzmann Transport Properties (BoltzTraP). The Seebeck coefficient of PbTe is higher than that of the other compounds examined at room temperature. Another thermoelectric attribute of SnSe is its increased electrical conductivity. At room temperature, $SnSe_2$ and SnSe have larger power factors. Spin-orbit coupling substantially influences these compounds' electrical and thermoelectric properties [37].

Trial-and-error material and synthesis condition selection often misses high-performance thermoelectric materials. Thus, the theory-guided search for high-performance thermoelectric materials with good electrical and thermal transport is growing [38].

NASA space missions use radioisotope power systems (RPS) for stable power. RPS generate electricity for spacecraft and sensors from plutonium-238 oxide decay heat. The current flight-proven RPS, the Multi-Mission Radioisotope Thermoelectric Generator (MMRTG), generates electricity from lead-telluride/tellurium-antimony-germanium-silver alloys. NASA SMD is investing in technology to improve Radioisotope Thermoelectric Generator (RTG) performance, including thermal-to-electric conversion efficiency (~6.3% at BOL) and deterioration rate over a 17-year lifespan.

NASA SMD invested in thermoelectric materials and device development to advance technology. Future RTGs could use high-efficiency materials found by researchers. SKD, Zintl phases, and lanthanum telluride (La3–xTe4) exhibit complex crystal structures, remarkable thermal and chemical stability at very low temperatures (1273 K to 473 K), and tremendous performance potential. In 2016, new thermocouple devices using these materials doubled conversion efficiency over outdated technologies.

Lower-temperature (900 K) SKD materials are being developed for an Enhanced Multi-Mission Radioisotope Thermoelectric Generator. The Radioisotope Thermoelectric Generator (eMMRTG) would enhance power by 25% at BOL and at least 50% at end-of-design life (17 years after BOL) due to its lower degradation rate. Skutterudite technology and higher-temperature 14-1-11 Zintl and La3–xTe4 materials could improve RTG designs. Segmented elements combine multiple materials into a single, wide-temperature, high-efficiency device that employs each material at its best. These modular designs could handle power systems from tens to hundreds of watts and have far higher specific power than the MMRTG. SMD-funded technology is also being examined for land-based industrial waste heat recovery and energy efficiency [39].

Green energy conversion technology allows waste heat to be used to cool spots with thermoelectric devices. However, most thermoelectric devices use toxic and expensive materials, restricting their use. These materials cannot be made on flexible, biodegradable substrates due to high temperatures. Many biocompatible and conformable devices are made using printed electronics.

Printing thermoelectric devices customized for individual purposes is cheaper and easier. Polymers and other bio-friendly materials are used to make inexpensive, lightweight, and flexible thermoelectric devices. Fortunately, low-temperature printing can produce some of these materials on eco-friendly surfaces.

Thermoelectric (TE) potential alleviates two main electronic system design challenges. Using waste heat, thermoelectric generators (TEGs) can power devices without the need for chemical energy storage. Microthermoelectric coolers can cool power-dense IC hotspots. Nuclear reactors generate less heat than high-performance CPUs, and good heat extraction improves chip performance and durability.

TEGs help wearable electronics and sensors absorb energy from body heat. This industry is underutilized due to inefficiency, toxicity, and high material costs. Traditional macro-scale TEGs are made from ingots divided into mm pallets, electrically connected with metal electrodes, and sandwiched between ceramic plates. This process is unsuitable for flexible substrates due to high temperatures and expensive equipment. Conventional technologies process only particular materials.

Printing and additive manufacturing are emerging flexible gadget production methods. Researchers have printed LEDs, solar cells, memristors, thin-film transistors, and artificial neural networks. Advantages include low-temperature, vacuum-less printing, cheap equipment, less waste, and flexible substrates. Technology for mass-produced electronics printing has been extensively studied. However, current printing technologies' low resolution, the substrate's roughness, and uneven layer thickness all reduce transistor performance.

TEGs have lower resolution requirements than transistors, whose gate length is inversely proportional to switching speed, frequency, and power gain. TEGs are usually in the mm-to-cm range, making printing an attractive alternative. Bulk devices function better than printed ones. Opposing performance characteristics are optimized in thermoelectricity. Ink and printing must be tuned for thermoelectrics [40].

Recent thermoelectric research has created new non-ceramic materials with advantages in elemental abundance, processability, cost, performance, and toxicity. These materials are partially investigated for thermoelectric applications but are not yet ready for widespread technology use. This field explores organic, carbon, or hybrid materials. The complex nanoscale design of most of these materials makes their thermoelectric performance limits potentially unknown. These materials focus on optimizing morphology and its use in devices to enable high ZT through new physics derived from complicated morphologies.

Unfortunately, long-term stability, device interface understanding, and cost-effective scale-up production prevent these materials from being widely used. With limited technological maturity, devices made from these materials are unlikely to be deployed soon but could have a long-term impact. Using thermoelectric materials can help achieve the government's zero-carbon goals. A semiconductor thermoelectric module converts electrical current into temperature gradients and vice versa. These modules provide electric heating and cooling without moving parts or refrigerants, and they can generate electricity from waste heat produced by other activities.

These devices are robust in aircraft and refrigeration applications. However, obstacles prevent the wider use of this technique. The low efficiency of current thermoelectric devices and the challenge of sustainable materials are serious issues. This research comes at a time when the thermoelectric technology market is growing, and new materials are being discovered. It is crucial to leverage this momentum to advance thermoelectric technology in society.

REFERENCES

1. Li, J. F., Liu, W. S., Zhao, L. D., & Zhou, M. (2010). High-performance nanostructured thermoelectric materials. *NPG Asia Materials*, *2*, 152–158. https://doi.org/10.1038/asiamat.2010.138
2. Ismail, B., & Ahmed, W. (2010). Thermoelectric power generation using waste-heat energy as an alternative green technology. *Recent Patents on Electrical Engineering*, *2*, https://doi.org/10.2174/1874476110902010027
3. Azam, S., & Ijaz, M. F. (2023). Chapter 4: Organic thermoelectric materials. In Ismail, B. I. A. (ed.), *New Materials and Devices for Thermoelectric Power Generation*. IntechOpen. https://doi.org/10.5772/intechopen.1002446
4. Petsagkourakis, I., Tybrandt, K., Crispin, X., Ohkubo, I., Satoh, N., & Mori, T. (2018). Thermoelectric materials and applications for energy harvesting power generation. *Science and Technology of Advanced Materials*, *19*, 836–862. https://doi.org/10.1080/14686996.2018.1530938
5. Yee, S. K., Coates, N. E., Majumdar, A., Urban, J. J., & Segalman, R. A. (2013). Thermoelectric power factor optimization in PEDOT:PSS tellurium nanowire hybrid composites. *Physical Chemistry Chemical Physics*, *15*, 4024–4032. https://doi.org/10.1039/C3CP44558E
6. Tritt, T. M. (2002). Thermoelectric materials: principles, structure, properties, and applications. In *Encyclopedia of Materials: Science and Technology* (Second Edition, pp. 1–11). https://doi.org/10.1016/B0-08-043152-6/01822-2
7. Mori, T., & Maignan, A. (2021). Thermoelectric materials developments: past, present, and future. *Science and Technology of Advanced Materials*, *22*, 998–999. https://doi.org/10.1080/14686996.2021.1966242
8. Zhang, Y., & Park, S. J. (2019). Flexible organic thermoelectric materials and devices for wearable green energy harvesting. *Polymers (Basel)*, *11*. https://doi.org/10.3390/polym11050909
9. Chang, W. B., Fang, H., Liu, J., Evans, C. M., Russ, B., Popere, … & Segalman, R. A. (2016). Electrochemical effects in thermoelectric polymers. *ACS Macro Letters*, *5*, 455–459. https://doi.org/10.1021/acsmacrolett.6b00054
10. Lv, A., Puniredd, S. R., Zhang, J., Li, Z., Zhu, H., Jiang, W., … & Wang, Z. (2012). High mobility, air stable, organic single crystal transistors of an n-type diperylene bisimide. *Advanced Materials*, *24*, 2626–2630. https://doi.org/10.1002/adma.201104987
11. Toshima, N. (2017). Recent progress of organic and hybrid thermoelectric materials. *Synthetic Metals*, *225*, 3–21. https://doi.org/10.1016/j.synthmet.2016.12.017.
12. Bubnova, O., & Crispin, X. (2012). Towards polymer-based organic thermoelectric generators. *Energy & Environmental Science*, *5*, 9345–9362. https://doi.org/10.1039/C2EE22777K
13. Hwang, S., Potscavage, W. J., Yang, Y. S., Park, I. S., Matsushima, T., & Adachi, C. (2016). Solution-processed organic thermoelectric materials exhibiting doping-concentration-dependent polarity. *Physical Chemistry Chemical Physics*, *18*, 29199–29207. https://doi.org/10.1039/C6CP04572C

14. Yang, Y., Deng, H., & Fu, Q. (2020). Recent progress on PEDOT:PSS based polymer blends and composites for flexible electronics and thermoelectric devices. *Materials Chemistry Frontiers*, *4*, 3130–3152. https://doi.org/10.1039/D0QM00308E
15. Atoyo, J., Burton, M. R., McGettrick, J., & Carnie, M. J. (2020). Enhanced electrical conductivity and seebeck coefficient in PEDOT:PSS via a two-step ionic liquid and NaBH4 treatment for organic thermoelectrics. *Polymers (Basel)*, *12*, 559. https://doi.org/10.3390/polym12030559
16. Qin, D., Cui, B., Yin, L., Zhao, X., Zhang, Q., Cao, J., … & Sui, J. (2019). Tin acceptor doping enhanced thermoelectric performance of n-type yb single-filled skutterudites via reduced electronic thermal conductivity. *ACS Applied Materials & Interfaces*, *11*, 25133–25139. https://doi.org/10.1021/acsami.9b05243
17. Qin, D., Shi, W., Lu, Y., Cai, W., & Sui, J. (2022). Efficient Si doping promoting thermoelectric performance of Yb-Filled $CoSb_3$-based skutterudites. *ACS Applied Materials & Interfaces*, *14*, 30901–30906. https://doi.org/10.1021/acsami.2c07044
18. Qiu, P., Shi, X., Chen, X., Huang, X., Liu, R., & Chen, L. (2011). Effects of Sn-doping on the electrical and thermal transport properties of p-type cerium filled skutterudites. *Journal of Alloys and Compounds*, *509*, 1101–1105. https://doi.org/10.1016/j.jallcom.2010.09.162
19. Latronico, G., Mele, P., Artini, C., Manfrinetti, P., Pan, S. W., Kawamura, Y., Sekine, C., Singh, S., Takeuchi, T., Baba, T., Bourgès, C., & Mori, T. (2021). Investigation on the power factor of skutterudite Smy $(Fe_xNi_{1-x})_4Sb_{12}$ thin films: effects of deposition and annealing temperature. *Materials*, *14*, 5773. https://doi.org/10.3390/ma14195773
20. Snyder, G. J., & Toberer, E. S. (2008). Complex thermoelectric materials. *Nature Materials*, *7*, 105–114. https://doi.org/10.1038/nmat2090
21. Ravindra, N. M., Jariwala, B., Bañobre, A., & Maske, A. (Eds.). (2019). Thermoelectric properties of silicon-germanium alloys BT - thermoelectrics: fundamentals, materials selection, properties, and performance. In *Thermoelectrics* (pp. 49–67). Springer International Publishing. https://doi.org/10.1007/978-3-319-96341-9_4
22. Huang, S., Ning, S., & Xiong, R. (2022). First-principles study of silicon–tin alloys as a high-temperature thermoelectric material. *Materials*, *15*. https://doi.org/10.3390/ma15124107
23. Rezania, A., & Rosendahl, L. A. (2017). Feasibility and parametric evaluation of hybrid concentrated photovoltaic-thermoelectric system. *Applied Energy*, *187*, 380–389, https://doi.org/10.1016/j.apenergy.2016.11.064
24. Su, X., Wei, P., Li, H., Liu, W., Yan, Y., Li, P., ... & Uher, C. (2017). Multi-scale microstructural thermoelectric materials: transport behavior, non-equilibrium preparation, and applications. *Advanced Materials*, *29*, 1602013. https://doi.org/10.1002/adma.201602013
25. Zhang, Z., Xie, Y., Peng, Q., & Chen, Y. (2016). A theoretical prediction of super high-performance thermoelectric materials based on MoS_2/WS_2 hybrid nanoribbons. *Scientific Reports*, *6*, 21639. https://doi.org/10.1038/srep21639
26. Liu, W. D., Yu, Y., Dargusch, M., Liu, Q., & Chen, Z. G. (2021). Carbon allotrope hybrids advance thermoelectric development and applications. *Renewable and Sustainable Energy Reviews*, *141*, 110800. https://doi.org/10.1016/j.rser.2021.110800
27. Bisht, N., More, P., Khanna, P. K., Abolhassani, R., Mishra, Y. K., & Madsen, M. (2021). Progress of hybrid nanocomposite materials for thermoelectric applications. *Advanced Materials*, *2*, 1927–1956. https://doi.org/10.1039/D0MA01030H
28. Ismail, B. I. A. (2023). New materials and devices for thermoelectric power generation. In Abed Ismail, B. (ed.). *IntechOpen* (pp. 1–152). https://doi.org/10.5772/intechopen.110935.

29. Yang, J., Yip, H. L., & Jen, A. K. Y. (2013). Rational design of advanced thermoelectric materials. *Advanced Energy Materials*, *3*, 549–565. https://doi.org/10.1002/aenm.201200514
30. Zhou, C., & Chung, I. (2020). Nanoscale defect structures advancing high performance n-type PbSe thermoelectrics. *Coordination Chemistry Reviews*, *421*, 213437. https://doi.org/10.1016/j.ccr.2020.213437
31. Ibáñez, M., Luo, Z., Genç, A., Piveteau, L., Ortega, S., Cadavid, D., ... & Cabot, A. (2016). High-performance thermoelectric nanocomposites from nanocrystal building blocks. *Nature Communications*, *7*, 10766. https://doi.org/10.1038/ncomms10766
32. Zhang, T. (2023). Advanced nanoscale materials for thermoelectric applications. *Nanomaterials*, *13*, 3615. https://doi.org/10.3390/nano13243165.
33. Li, H., Zhang, C., Li, P., Liu, S., Zhang, H., & He, C. (2023). Recent development in flexible organic thermoelectric fibers for wearable devices. *Materials Today Chemistry*, *34*, 101774. https://doi.org/10.1016/j.mtchem.2023.101774
34. Shi, X., Chen, L., & Uher, C. (2016). Recent advances in high-performance bulk thermoelectric materials. *International Materials Reviews*, *61*, 379–415. https://doi.org/10.1080/09506608.2016.1183075
35. Mao, J., Liu, Z., Zhou, J., Zhu, H., Zhang, Q., Chen, G., & Ren, Z. (2018). Advances in thermoelectrics. *Advances in Physics*, *67*, 69–147. https://doi.org/10.1080/00018732.2018.1551715
36. Muhtaroglu, A., von Jouanne, A., & Yokochi, A. (2007). Hybrid thermoelectric conversion for enhanced efficiency in mobile platforms. *Journal of Micromechanics and Microengineering*, *17*, 1767. https://doi.org/10.1088/0960-1317/17/9/003
37. Khan, A. A., Khan, I., Ahmad, I., & Ali, Z. (2016). Thermoelectric studies of IV–VI semiconductors for renewable energy resources. *Materials Science in Semiconductor Processing*, *48*, 85–94.
38. Gutiérrez Moreno, J. J., Cao, J., Fronzi, M., & Assadi, M. H. N. (2020). A review of recent progress in thermoelectric materials through computational methods. *Materials for Renewable and Sustainable Energy*, *9*(3), 16.
39. https://smd-cms.nasa.gov/wp-content/uploads/2023/04/NASA_2016SMD_Report_final_highres.pdf
40. Hossain, M. S., Li, T., Yu, Y., Yong, J., Bahk, J. H., & Skafidas, E. (2020). Recent advances in printable thermoelectric devices: materials, printing techniques, and applications. *RSC Advances*, *10*(14), 8421–8434.

8 Challenges and Opportunities for Thermoelectric Materials and Devices

8.1 MATERIALS SCALABILITY AND COST CONSIDERATIONS

Thermoelectric (TE) materials convert heat into sustainable energy. The devices are eco-friendly, low-maintenance, and have no moving parts, but inefficient materials and thermal stability limit TE efficiency. High-pressure synthesis (HPS) and spark pressure synthesis (SPS) processes for thermoelectric measurements demonstrate their pros and cons. Large-scale thermoelectric device manufacturing benefits from HPS's industrial scalability and adaptability. Faster processing and longer cycles may increase grain growth and lower thermoelectric efficiency. SPS is popular due to its quick processing, precise temperature control, and low, homogeneous microstructure. Global energy consumption is highest in transportation and industry, which use fossil fuels and generate tremendous heat. Energy reuse minimizes CO_2 emissions, global warming, and fossil fuel consumption.

Optimizing thermoelectric material synthesis and customizing fabrication techniques for different thermoelectric applications requires understanding the benefits and downsides of each methodology. This expertise helps select sintering methods based on thermoelectric device performance and production scalability [1]. SPS and HPS inorganic thermoelectrics are more efficient than 3D printing. High binder content makes part densification difficult in direct-ink-writing manufacturing. Sintering for long periods and adding filler and sintering aids to the ink can produce high-density printed thermoelectrics [2].

Wireless sensor networks (WSNs), as well as laptops, cellphones, and tablets, are the most common TEG applications in microelectronics. Defining "microelectronics" as any technology that incorporates micro components broadens this application range. Thin films are the most studied technology, since micro-TEG allocations are modest. Melting mixed raw materials to generate polycrystalline powders for bulk TEMs is still popular due to its productivity, scalability, low cost, and ease of use. Due to superior product quality, the ability to deposit any material, and scalability, small-scale TEM production through evaporation or deposition appears promising [3].

Modules can be merged to create larger thermoelectric generators. Different thermoelectric generator designs are needed for various applications due to

DOI: 10.1201/9781003639343-8

operating variables such as temperature and heat source intensity. Since waste heat is free, TEGs' compactness, scalability, and dependability won't improve the efficiency of these applications. Scalable and small TEGs can recover heat from multiple sources. Thermoregulators simplify waste heat recovery processes, such as metal casting [4].

These processing parameters scatter phonons at grain and domain boundaries, lowering thermal conductivity (k) without impacting the Seebeck coefficient (s) and enhancing the figure of merit (ZT). Due to metal shortages and toxicity, silicon and transition-metal oxides have replaced Te- and Pb-based materials for scalability and safety. Pure CNTs are expensive and unscalable for TE applications; hence, polymer composites incorporating CNTs are preferred [5].

8.2 COST CONSIDERATION

Waste-heat recovery and targeted cooling thermoelectrics are becoming more widespread despite cost and scalability issues. Material, production, and device analysis affect costs. The heat exchanger cost, C, depends on factors such as heat transfer coefficient, U, active material density, ρ, leg length, and ceramic plate area, A [6]:

$$C = \left(\left(C_B + C_{M,B}\right)\rho L + C_{M,A}\right)AF + C_{HX}UA \tag{8.1}$$

and

$$C''' = \left(C_B + C_{M,B}\right)\rho L \tag{8.2}$$

$$C'' = C_{M,A} \tag{8.3}$$

First, examine volumetric cost, C, and then areal cost, C, while considering manufacturing costs. Future product costs, rather than device component prices, are considered. Devices on the market will benefit from rapid cost estimate iterations throughout thermoelectric product development. The extractive raw bulk material cost, C_B, reflects material cost differences. Base components are usually 99% pure for thermoelectric materials. Most minerals require purification; hence, the cost-performance analysis does not include purification costs.

Comparisons of pure and raw material costs did not influence material rankings. There are two material processing manufacturing cost categories. Cost and throughput estimates were provided by most equipment providers. Depending on the material and application, throughput cannot be standardized. A year-round activity measure is used for this industrial equipment. Manufacturing costs are limited by equipment efficiency. Repairs and maintenance might reduce equipment efficiency. The realism of equipment effectiveness has little effect on these results. Higher-yield thermopower materials require fewer legs and less processing. The active material may differ from the device due to production losses.

Manufacturing-related material loss is ignored to calculate a cost lower bound. New manufacturing equipment costs are expected. Equipment is frequently designed for the end-use application; thus, vendors must be contacted for exact quotes and costs. Service providers or user facilities may be cheaper than owning equipment and in-house manufacturing. Equipment and production costs serve as references; material, technique, and application determine manufacturing costs.

Chalcogenides include silicides, clathrates, half-Houser alloys, skutterudites, and oxides. Screen-printed polymers have special production costs. Alternatives to dicing include printed metallization layers. Fabricators classify materials as bulk, nanobulk, nanowire, superlattice, and polymer. The cost-performance optimum leg length, outside manufacturing constraints, requires new high-cost production methods. DICING saw instructions recommend a 500 mm leg length for bulk materials [6].

The cost-performance ratio drives technology adoption, not performance. PV cells have become popular due to their cost-effectiveness of $1/watt. Similarly, TEGs should be assessed like other energy sources. The four TEG module combinations have two TE-made p- and n-type segmented TEG legs. Using type II BCs with a constant hot side temperature (1000 K) and variable cold side heat transfer coefficients based on cross-sectional area ratios yielded the optimum cost-performance TEG leg lengths. When price is less important, TE materials with high power factors and low ZTs may be useful.

The capital costs include

i. TE material costs,
ii. TE module production costs, and
iii. Heat exchanger costs on both sides.

Gravimetric and areal components constitute manufacturing costs. Hot pressing, ball milling, etc. depend on TE material weight. Therefore, the areal cost of this segmented TEG leg is about N times the leg's cross-sectional area plus the manufacturing price. Operating costs are excluded because this cost-performance ratio compares overnight capital cost to TEG output power [7].

Device performance and commercial viability depend on fabrication costs and the thermal and electrical transport characteristics of prospective thermoelectric materials. High cost and low conversion efficiency limit TEGs' applications in medicine, military, and space. Since kerosene lamps, thermoelectric generators have been employed in aerospace, industrial utilities, transportation, medical services, electronics, and temperature measurements. Thermoelectric devices serve as coolers, power generators, and thermal energy sensors in military, aircraft, instrumentation, biology, medical, and industrial applications. Besides photovoltaics, thermoelectric generators may also generate power from solar energy. System recovery takes 6–8 years, depending on thermoelectric generator prices. Space missions and isolated areas benefit from thermoelectric technology, such as powering fans and illumination with wood burner thermoelectric generators [8].

8.3 INTEGRATION WITH ENERGY EFFICIENCY TECHNOLOGIES

Fin architecture, size, and weight limit thermal transport even in the largest copper heat sinks. Contouring, grooves, perforations, and fluid flow have been used in many heat sink designs. Specific nanofluids have been researched to enhance TEG performance. Other research optimized heat sink perforations for TEG system convective heat transfer using numerical and experimental methodologies. Using diverse methods, carbon nanotube (CNT) integration on copper faces many obstacles. The quality and structure of copper-grown, water-vapor-assisted CVD CNTs are unclear. Thermal chemical vapor deposition (CVD) and microwave plasma sputter-deposition of metal catalyst films have deposited multiwalled CNTs on copper; however, their orientations and repeatability are unknown.

With various heat sinks, Figure 8.1 shows the TEG open-circuit output voltage versus time. The heater is set to 100°C, and the ambient temperature is maintained at 21°C for the studies. A simple copper sheet reduces the TEG output voltage by 7%, lowering cool-side heat transfer. Importantly, CNT-covered copper sheets increase the output voltage above that of finned copper heat sinks [9].

First-generation thermoelectric materials included bismuth telluride. Thermoelectric material research is evolving and diverse, as these materials may increase performance and efficiency in various applications. TEG research continues to focus on energy fluxes, heat transport, and system development. Novel geometries and nanostructured materials are being explored to improve TEG efficiency [10].

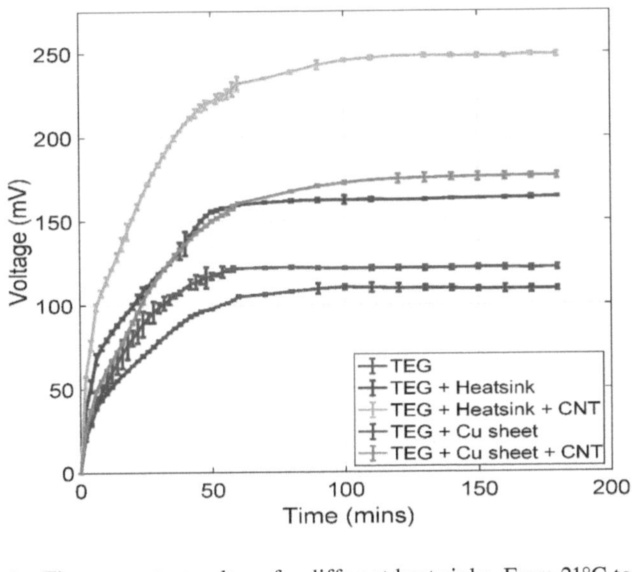

FIGURE 8.1 Time vs. output voltage for different heat sinks. From 21°C to 100°C, the heater is heated [9].

8.4 EMERGING THERMOELECTRIC MATERIALS AND DEVICES

Transverse thermoelectricity (TTE) uses the Nernst effect to harvest energy from surroundings instead of Seebeck-based longitudinal thermoelectric materials. Recently, material exploration and performance-boosting strategies have advanced with new technologies. Unlike thermoelectric semiconductors, TTE materials are topological semimetals with high carrier mobility. Instead of the bipolar effect in conventional TE, a magnetic field offers the synergistic thermopower effect.

Electronic and hole carriers travel in opposite directions. However, longitudinal temperature gradients produce transverse thermopower. Because transport directions are perpendicular, electrical and thermal conductivity can be adjusted separately. In the Nernst configuration, the bipolar effect and Wiedemann–Franz law that limit Seebeck effect-based thermoelectric devices are abolished.

Finding topological semimetals advances TTE material research. The degeneracy of energy band nodes and momentum space distribution distinguishes Dirac, Weyl, and Nodal-line topological semimetals. By creating nonzero $\partial V/\partial r$ on the y-axis, $\partial T/\partial r$ on the x-axis, and B on the z-axis, the Nernst effect-based transverse thermoelectric (TTE) addresses the concerns above. The magnetic field provides a new degree of freedom, the synergistic thermopower effect, instead of the bipolar effect in conventional TE. Electrons and holes move oppositely. Transverse thermopower comes from longitudinal temperature gradients. Electrical and thermal conductivity can be changed separately since transport directions are perpendicular.

The Nernst configuration reduces the bipolar effect and Wiedemann–Franz law in Seebeck effect thermoelectric devices. TTE materials carry electrons and holes due to electron-hole compensation. The P- and N-type materials can be replaced in TTE devices. Module manufacturing is efficient due to transverse geometry and high TTE. Bismuth-incompatible materials prevented the TTE effect from being discovered for a century. Its narrow Fermi surface, carrier mobility, and electron-hole correction give bismuth a significant TTE effect. TTE material research advances from topological semimetals. Topological semimetals, like bismuth, have a band structure with a lowered Fermi surface, a 0 eV energy gap, and linear dispersion, all of which satisfy the high TTE effect requirements. The energy band nodes and momentum space distribution of Dirac, Weyl, and Nodal-line topological semimetals are degenerate (Figure 8.2) [11].

Researchers found that grain shape, interface density, and orientation improved TE performance by reducing thermal conductivity and enhancing electrical characteristics by reducing secondary phases.

Nanoscale barriers, inclusions, and composites direct low-energy carriers to high-frequency/energy carriers (Figure 8.3). By adding nano-precipitates to the matrix, TE materials improve their power factor by increasing S and σ. Recently, modulation doping has optimized TE in $Ti_{(1-x)}Mn_xCoSb$ [12].

Due to large grain sizes, nano-scale size reduction was examined to reduce heat conductivity. Nanoparticle surface and contact regions increase phonon

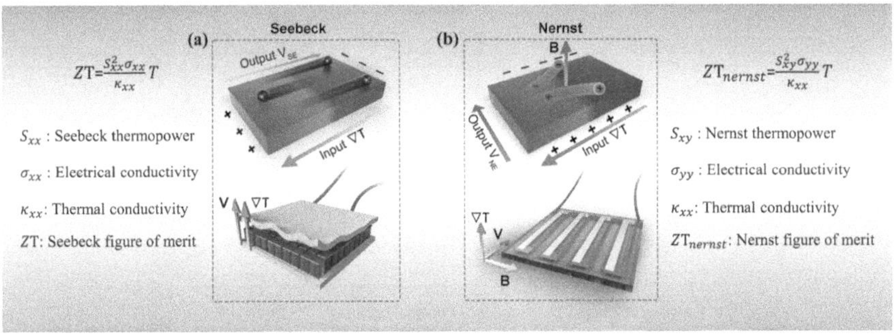

FIGURE 8.2 Schematics of (a) Seebeck effect and LTE and (b) Nernst effect and TTE devices [11].

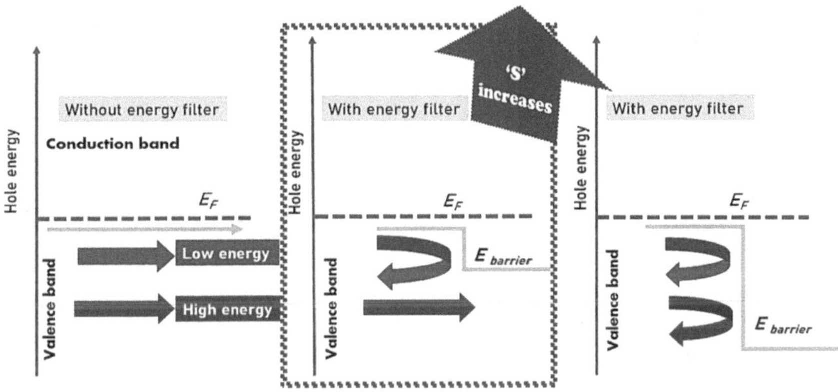

FIGURE 8.3 Energy filtering schematic using nano-composites and nano-inclusions [12].

scattering, lowering thermal conductivity. Acid treatment improved grain boundaries. Treatments included H_2SO_4 and HCl. Figure 8.4 shows HCl-treated NbFeSb FE-SEM images. Small, irregular grain size reduction by HCl was predominantly around particle edges. In contrast, H_2SO_4 reduced particle size (Figure 8.5a and b).

Clean and acid-treated NbFeSb grains have become sub-50 nm particles in FE-SEM images (Figure 8.4d). Different acid concentrations—30% HCl (12 M) and 98% H_2SO_4 (18 M)—may explain this difference. Acid concentration impacts grain size and uniformity. High H_2SO_4 levels could cause particle oxidation after acid treatment. NbFeSb and At-NbFeSb XRD patterns are shown in Figure 8.5c. XRD showed no oxidation peaks following high-concentration acid treatment. Nanoparticle size affected XRD peaks. Figure 8.5d shows the highest intensity of high-resolution XRD peaks in the 42°–44° region. When XRD patterns were normalized using the highest intensity peak, peak shifts and lower FWHM were observed.

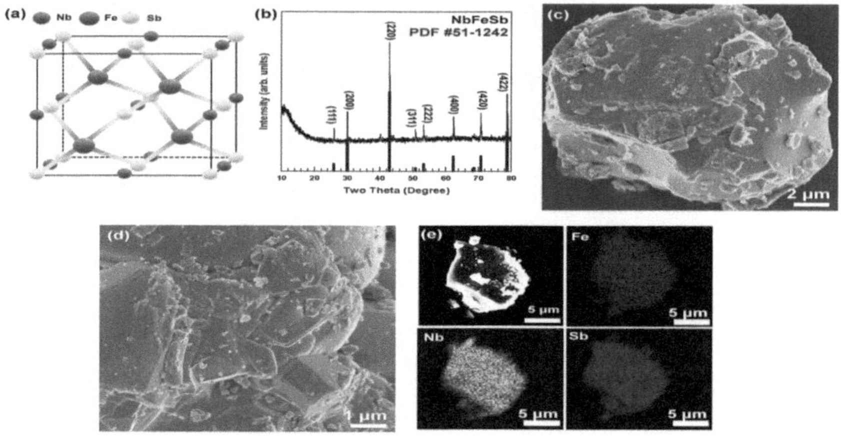

FIGURE 8.4 (a) NbFeSb half-Heusler cubic crystal structure. XRD pattern of NbFeSb. (c) Fe-SEM picture and (d) high-resolution NbFeSb image. (e) EDS elemental mapping verified Nb, Fe, and Sb atoms in NbFeSb [13].

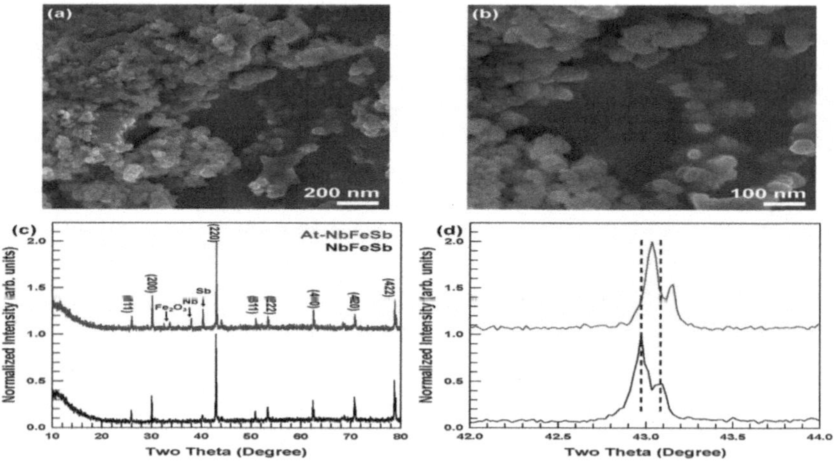

FIGURE 8.5 (a and b) At-NbFeSb FE-SEM pictures. (c and d) XRD patterns of At-NbFeSb and NbFeSb at high resolution [13].

Treatment with acid increased FWHM and XRD peak angles. NbFeSb morphology changed dramatically following acid treatment. Acid treatment corroded the perfect surface, creating several defects. Additionally, particle sizes were drastically reduced to nanometers. FE-SEM images match these results in Figure 8.5a and b. NbFeSb and At-NbFeSb XPS scans and high-resolution Sb 3d peaks are shown in Figure 8.5. After acid treatment, Sb 3d peaks revealed moderate changes and no new peaks, confirming a 530 eV oxygen deficit. This shows that H_2SO_4-treated NbFeSb reduced grain size without oxidation.

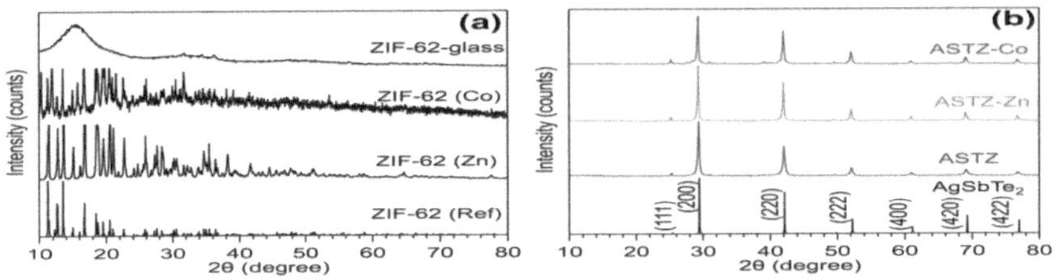

FIGURE 8.6 (a) Synthesized crystalline and amorphous ZIF-62(Zn/Co) and (b) ASTZ, ASTZ-Zn, ASTZ-Co PXRD patterns [15].

Resistivity and Seebeck coefficient define thermoelectrics. These measurements indirectly measure microscopic carrier concentration and mobility. Use the electrical structure-sensitive Seebeck coefficient to analyze microscopic and transport aspects. The Seebeck coefficient of metallic Cu_2Se displays a unique temperature dependency at 350 K, indicating phase transitions and two-phase transitions. The Cu_2Se phase transition's kinetic limitation on the Seebeck coefficient can be shown by thermal study at different heating rates. The large Seebeck effect in metallic Cu_2Se may benefit thermoelectrics [14].

In Figure 8.6a, synthesized ZIF-62-Zn, ZIF-62-Co, and agZIF-62(Co/Zn) powder XRD patterns are shown (see Figure 8.6b) for ASTZ-Zn, ASTZ-Co, and ASTZ powder XRD patterns. ASTZ, ASTZ-Zn, and ASTZ-Co have identical XRD patterns, and ZIF-62's absence of peaks suggests that heat during synthesis makes it amorphous. The ASTZ-Zn and ASTZ-Co XRD data show more low-intensity peaks, indicating non-ZIF-62/ASTZ phase peaks. FTIR peaks of ASTZ (Co/Zn) match amorphous ZIF-62. After 823 K heat treatment, ASTZ (Co/Zn) exhibited amorphous ZIF-62 and negligible organic linker degradation. Rietveld refinement of ASTZ-Zn/Co reveals its crystal structure in Figure 8.7.

$Ag_{0.85}Sb0_{.8}Zn_{0.04}Te_2$ is the ASTZ sample's refined crystal lattice metal ratio. Extra ZIF-62(Zn/Co) peaks are not part of the ASTZ phase. When MOF is introduced to ASTZ, the pores may contain silver tellurides and Zn_4Sb_3 if Ag and Sb are displaced. Therefore, ASTZ-Zn's refined crystal lattice metal ratio is $Ag_{0.6}1Sb_{0.76}Zn_{0.31}Te_2$. Zn displacement and Ag_3Sb phase development predict decreasing Ag and Sb formula ratios and increasing Zn. The formula ratio for ASTZ-Co is $Ag_{0.68}Sb_{0.7}6Zn_{0.03}Te_2Co_{0.028}$, where Co replaces crystallographic Ag atoms. The peak at 39.31° represents the orthorhombic Ag_3Sb phase, while the peak at 40.19° mixes silver telluride and Ag3Sb phases Moieties in composites may interact to generate tiny peaks. PXRD phase studies and X-ray refinement show ASTZ's crystal structure interacts with Ag ZIF-62. This interaction alters the crystal lattice metal ratio and creates non-ASTZ phases in ASTZ-Zn or ASTZ-Co.

Temperature-dependent transport measurements reveal material system processes. A chip resistor and E-type thermocouple measured the sample's temperature differential. To test sample electrical conductivity (σ), the two-probe approach was used, which involves applying voltage and measuring current. This approach requires a sample temperature gradient and voltage difference measurement.

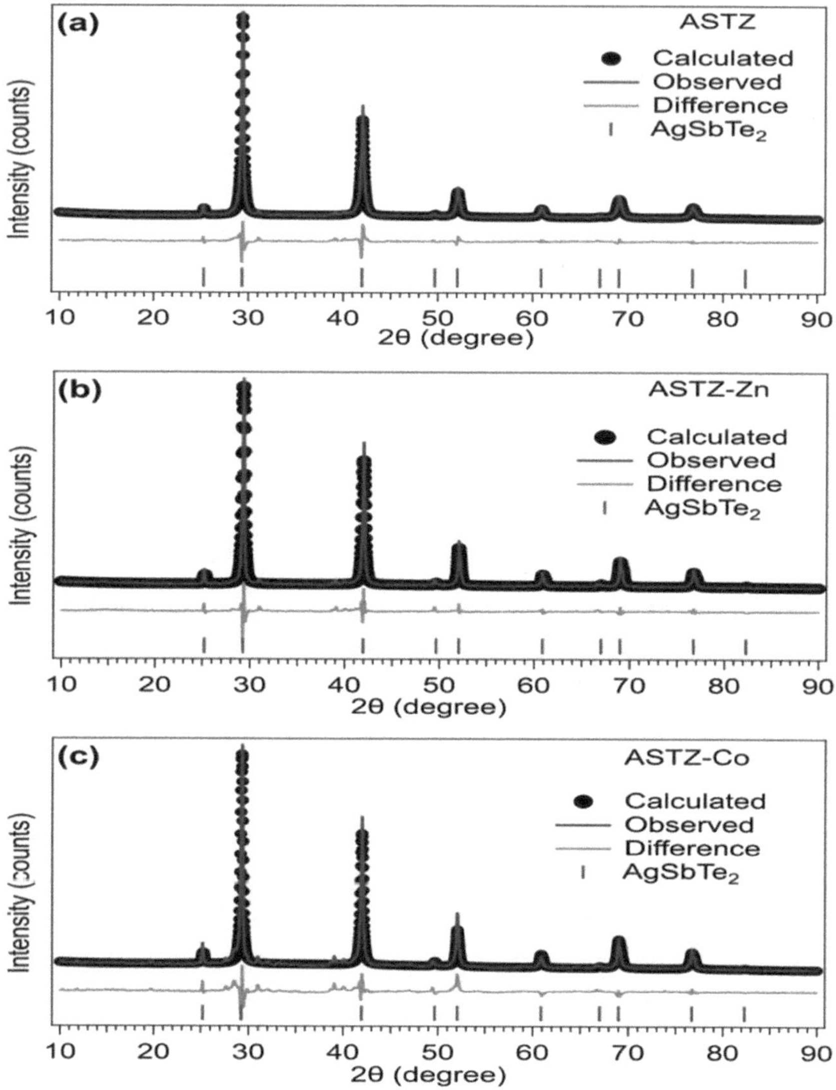

FIGURE 8.7 Refinement graph of (a) ASTZ, (b) ASTZ- Zn, and (c) ASTZ-Co [15].

A precise voltage measurement and temperature adjustment yielded the Seebeck coefficient. Measurements from room temperature to 400 K were conducted to investigate the samples' temperature-dependent thermoelectric responses. Both samples' Seebeck coefficients increased with temperature (Figure 8.8a). Materials with thermally generated carriers exhibit this property. As temperature increased, thermal energy added charge carriers, boosting the Seebeck coefficient.

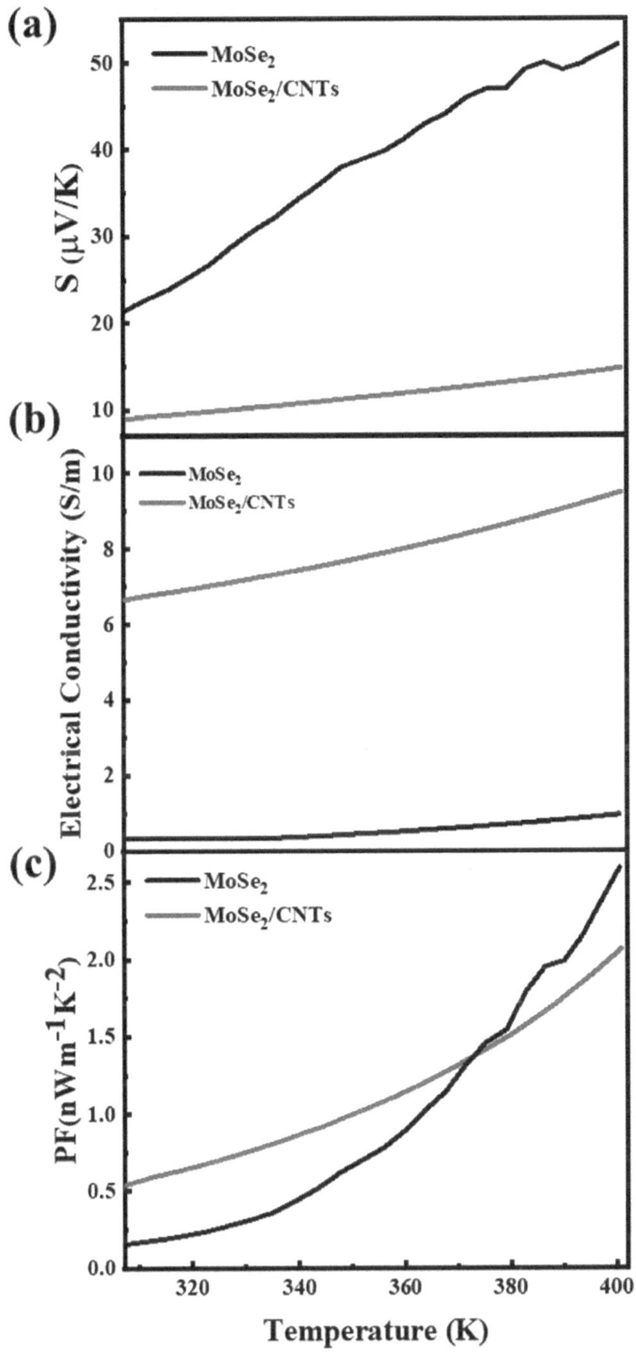

FIGURE 8.8 (a) The Seebeck coefficient, (b) electrical conductivity, and (c) power factor of $MoSe_2$ and its nanocomposite vary with temperature (305–400 K) [16].

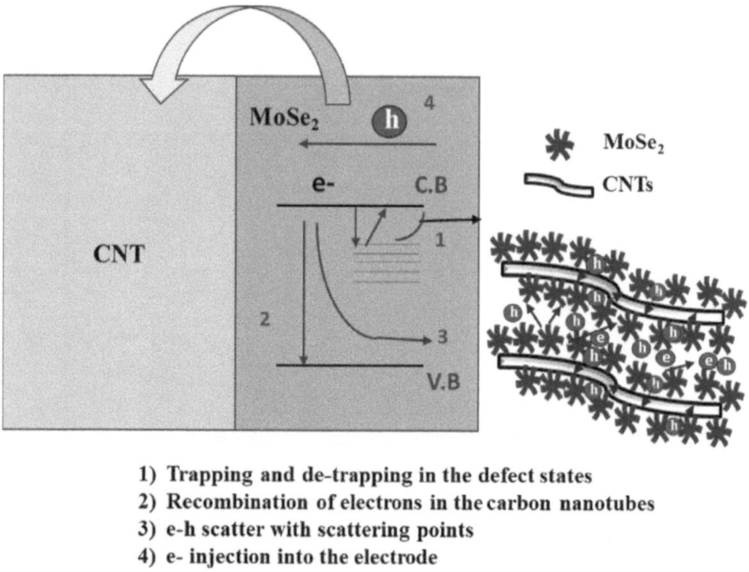

FIGURE 8.9 Schematic of CNT-$MoSe_2$ nanocomposite carrier transport [16].

$MoSe_2$'s Seebeck voltage was higher than the other sample's due to the nanocomposite's continuous CNTs. The high CNT surface area boosted the nanocomposite's phonon-carrying performance and minimized the temperature gradient. The EDS spectra of $MoSe_2$ and the CNT nanocomposite are shown. For p-type conductivity, Se's interstitial states offered an electron-accepting acceptor level near the valence band (Figure 8.8a). Temperature-induced CNT wall defects were increased by environmental oxygen doping and functionalization-attached hydroxyl and carboxyl groups. Figure 8.8b demonstrates that $MoSe_2$/CNTs had better conductivity than pure CNTs. Mobility boosters, like CNTs, boosted the nanocomposite sample conductivity. Charge carrier mobility and kinetic energy increased transfer and material electrical conductivity with CNTs. In the nanocomposite, CNTs bridged $MoSe_2$ layers to facilitate carrier migration (Figure 8.9). CNTs lower the Schottky barrier, increasing conductivity.

Material semiconductivity was confirmed by the temperature-induced conductivity rise. Figure 8.8c shows how CNTs alter the nanocomposite's power factor. Electrical conductivity and the Seebeck coefficient illuminated transport and material performance. It changed at higher temperatures. Both materials' Seebeck coefficients increased at 400 K, while the $MoSe_2$/CNT nanocomposite's power factor dropped from 2.59 to 2.06 nW/mK^2. Given its large aspect ratio, MWCNTs have quasi-ballistic electrical conductivity. $MoSe_2$ linked to additional nanotubes in the nanocomposite created charge carrier pathways. Nanocomposite conductivity and the $MoSe_2$ power factor rose. Anderson localization hindered carrier mobility in ideal $MoSe_2$, but CNTs enhanced nanocomposite carrier mobility.

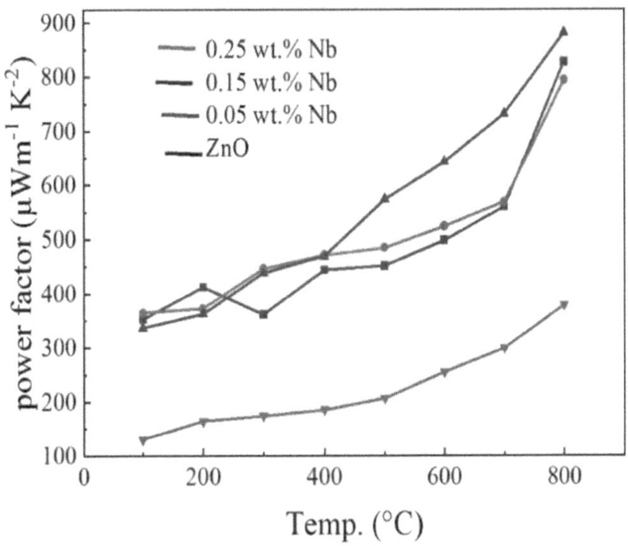

FIGURE 8.10 Temperature affects sintered NZO ceramic power factor [17].

Figure 8.10 shows the temperature-dependent power factor (P.F.) of NZO ceramics. Increasing temperature from 100°C to 800°C linearly increases the P.F. of all materials. The P.F. measures thermoelectric system performance by electricity. ZnO doped with 0.15 wt% Nb reached a maximum P.F. value of ~8.8×10^{-4} W/m K^2 at temperatures from 100°C to 800°C. These compositions have higher power factors than ZnO-based ceramics.

Technical applications require novel cubic spinel materials AB_2X_4 (where A and B are cations and X are chalcogenide anions) since their physical properties depend on A- and B-site distributions. Adjustable exchange splitting occurs due to the host lattice's ionic distributions affecting cationic and anionic energy states. Magnetic and thermoelectric chalcogenide spinels abound, being useful for energy conversion due to their optical and thermoelectric properties. $MgHo_2Z_4$ (Z=S, Se)'s half-metallic ferromagnetic and thermoelectric properties were examined using DFT. Spintronic and thermoelectric energy conversion technologies showed promise. Hypothetically, half-metallic ferromagnetic $MgTi_2S_4$ converts thermoelectric energy better than $MgMn_2S_4$.

Another DFT study found ferromagnetic semiconducting $MgPr_2X_4$ (X=S, Se) spinels that can create thermoelectric power. DFT simulations were utilized to study the spinels HgY_2S_4 and HgY_2Se_4, which have optical and transport properties appropriate for commercial optoelectronic and thermoelectric devices. Spinels are also proposed for Mg ion battery energy storage. Chalcogenide anions' tetrahedral surroundings have many lattice sites, most of which are empty, but a few are filled by Mg cations due to spinel structural composition. Mg^{2+} migrates across the lattice, suggesting Mg-based spinel halides could be solid electrolytes in energy-storing ion batteries.

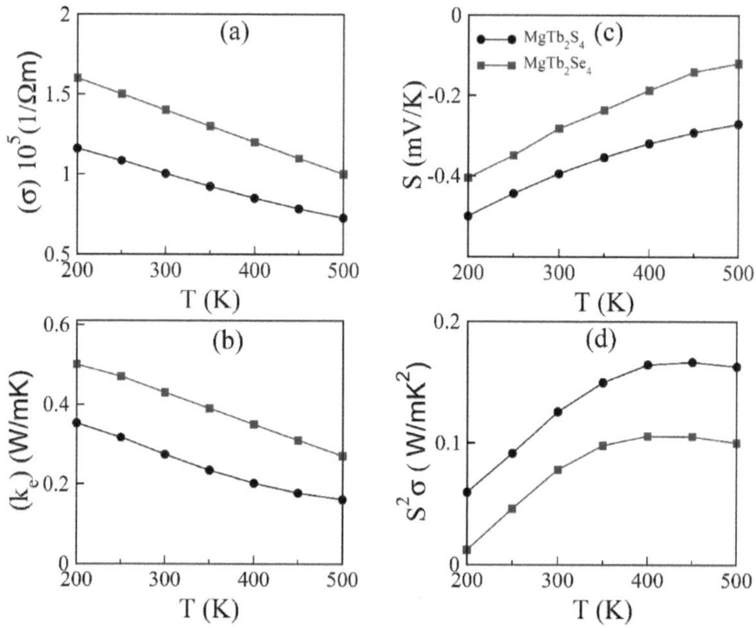

FIGURE 8.11 (a–d) Computed thermoelectric parameters (σ, κ, S, σS^2) for $MgTb_2$(S/Se) [18].

Figure 8.11 shows room-temperature thermoelectric properties for both spinels. Figure 8.11a shows semiconducting spinels at 200–500 K. At 300 K, $MgTb_2S_4$ and $MgTb_2Se_4$ had σ values of 1.0×10^5 (1/Ωm) and 2.36×10^5 (1/Ωm), respectively, indicating that replacing S with Se increases electron mobility. Electron and lattice contributions determine thermal conductivity κ. At ambient temperature, $MgTb_2S_4$ and $MgTb_2Se_4$ have κe values of 0.28 W/mK and 0.42 W/mK, respectively (Figure 8.11b). The Seebeck coefficient S measures metal-contact potential change. The studied materials' S rises with temperature (Figure 8.11c). The computed S reveals $MgTb_2Se_4$ is larger than $MgTb_2S_4$. The Seebeck coefficients for $MgTb_2S_4$ and $MgTb_2Se_4$ at 300 K are −0.40 and −0.30 mV/K, respectively. The power factor (σS^2) measures how successfully materials convert heat energy into electricity, which is electric conductivity times the Seebeck coefficient squared. In Figure 8.11d, both spinels' power factors are plotted against temperature. $MgTb_2S_4$ has a 0.13 W/mK² room-temperature power factor, while $MgTb_2Se_4$ has 0.08. $MgTb_2S8$ is better for thermoelectric devices due to its higher estimated ZT.

8.5 POTENTIAL IMPACT AND OPPORTUNITIES FOR THERMOELECTRIC MATERIALS AND DEVICES

Recently, some material systems have high ZT values, improving laboratory generation. Cheaper manufacture of these materials could lower the prices of high-efficiency thermoelectric waste heat recovery. Alternatives are needed since traditional materials have problems. Because of lead's regulatory restrictions, tellurium demand could change drastically depending on the CdTe solar cell market. Other alternate material classes' benefits are discussed. Tetrahedrites and magnesium silicides are cheaper, but half-Heusler materials can withstand high temperatures. Using bulk thermoelectric materials like Skutterudite and Half-Heusler, advanced TEGs are being designed, tested, and demonstrated on an engine dynamometer and full-scale vehicle to increase fuel economy by at least 5%. Improved production and cost reduction will make these updated TEGs commercially viable at 100,000 units per year [19].

The solid-state nature of thermoelectrics makes them compact, vibration-free, and durable. TE materials are used for cooling and waste heat conversion. Peltier coolers provide reliable temperature control in Charge Coupled Device (CCD) detectors and microprocessor heat sinks. Portable electronics, medical sensors, and combustion engine exhaust pipe modules use body heat. Commercialization requires energy conversion efficiency improvements for these generators [20].

Evaporation and latent heat from water are omnipresent. Hot water vapor powers a self-sustaining polymeric nanogenerator for energy recovery. In Figure 8.12a, water vapor drives a generator. Many water droplets heat the generator following water-vapor condensation. After water droplets absorb airflow heat, the generator cools quickly and humidity drops. Figure 8.12b and c shows a generator on hot coffee and water vapor driving blue LEDs. Peak output power against loading resistance for water-vapor-powered 28 mm PVDF and 25 mm P(VDF-TrFE) generators is shown in Figure 8.12d. Both vapor-driven generators' loading resistance rises and declines.

Due to their higher pyroelectric coefficient (26–27 mC/m²), PVDF generators provide more power than P(VDF-TrFE) generators. Figure 8.12e shows the use

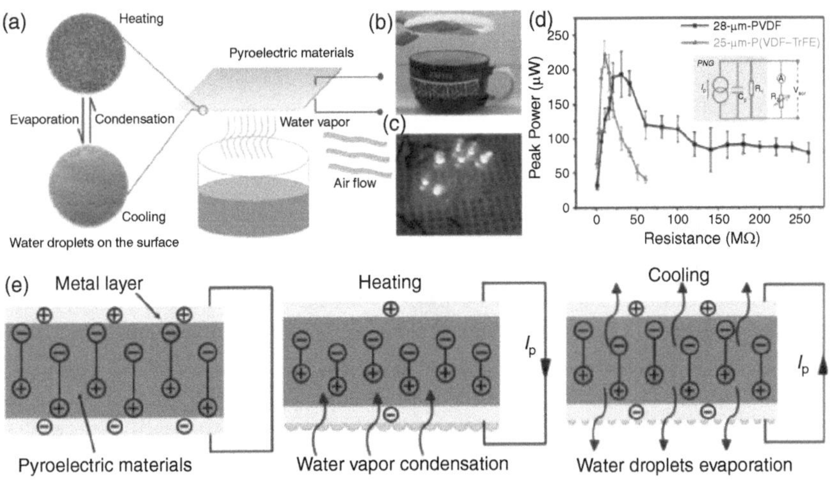

FIGURE 8.12 (a) Schematic of a water vapor-driven generator, (b) generator above hot coffee, (c) flashed blue LEDs powered by generator, (d) peak output power vs. loading resistance and inset circuit diagram, and (e) water vapor-driven generator mechanism [21].

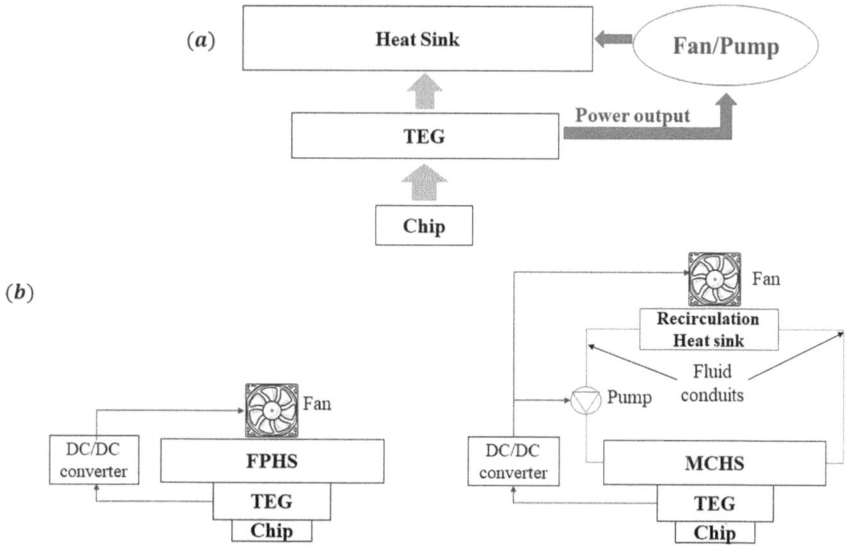

FIGURE 8.13 (a) A sample thermoelectric-self-cooling system and (b) a forced-air and water-cooling system [22].

of water vapor to produce power. The eternal dipole charges two metal electrodes perpendicular to polarized PVDF/P(VDF-TrFE) films. Higher generator temperatures reduce dipole moments and enhance water vapor condensation. Due to lower polarization density, an external circuit current balances the metal electrode charge density.

Figure 8.13a indicates that electric devices need cooling. A heat sink's cold side, integrated with a thermoelectric generator (TEG), could self-cool. The TEG powers the fan or pump that circulates coolant across the heat sink by converting electrical device heat into energy using the Seebeck effect. Powerless thermoelectric self-cooling systems can cool overheating equipment.

More recently, sophisticated manufacturing technologies can mimic nanoscale features that enhance ZT values in bulk materials. Nanostructure thermal stability is crucial for high-temperature power generation. Short-term thermal stability (grain formation and interdiffusion) is high for nanobulk $Si_{1-x}Ge_x$, despite experimental evidence to the contrary.

Scalable material synthesis through powder metallurgy methods and long-term thermal stability are needed. Nanobulk $Si_{1-x}Ge_x$ materials are suitable for high-temperature thermoelectric applications; however, results vary by class.

Two external cold-side exchangers and a core hot-liquid-metal heat exchanger are connected by thermoelectric couple arrays. These convert heat at hundreds of W/cm² for small thermoelectric converters. Thinner thermoelectric components, less than 5–10 mm, have heat flux densities of 1–40 W/cm², making thermal losses difficult to regulate. Highly corroded heat exchanger surfaces from waste heat gas streams make high-temperature heat transfer challenging.

For waste heat recovery, cold-side heat exchangers should use liquids where possible. Cold-side heat exchangers can employ proven technology and function at low temperatures because liquids have orders of magnitude superior thermal conductivities and capacities than gases [23].

TEG systems can replace batteries in many applications and are compact, inexpensive, and organized. The preamplifier and sensor control of a 1.5 mW TEG system with a 10°C temperature differential could be small.

Heating with any source uses little power. The Internet of Things (IoT) is a major technological innovation of this century that can provide various services. Sensors, computation, and networking technologies communicate to collect data in the IoT without human intervention. Using IoT in transportation, health, manufacturing, and education improves lives. A cost-effective and consistent power supply is needed to power IoT devices.

TEG is a renewable energy conversion technology for IoT modules that has been studied. Thermoelectric modules increase waste heat recovery and energy generation in cars. A revolutionary foldable thin-film TEG for large-area, low-energy-density applications has been developed. Innovative folding raised the temperature gradient and reduced heat loss.

TE systems can help reduce global energy consumption and pollution. Domestic AC and power generation can benefit from improved merit and arrangements. In compact instruments, thermoelectric cooling is superior to conventional cooling; hence, thermoelectric modules are used for self-cooling. Future cooling systems would benefit from TECs. Optimization of TE devices requires finite-time thermodynamics. Inserts in channel walls increased output power by 110%. Optimization improved TEG module output power from 1W to 9W. Temperature increases power production, but TEG durability and efficiency limit it. TEG systems can generate solar energy alongside PV panels. TEGs pay off in 6–8 years [24].

New applications need different materials and pose various challenges. Material stability's obvious challenge is rarely discussed. Over 200°C, most organic-based soft thermoelectrics deteriorate, restricting their use to low temperatures. To dope polymers, small compounds that are unstable to temperature and oxygen are utilized. Organic semiconductors degrade over 120°C. Small-scale and portable electronics have made thermoelectric power production and spot cooling applications popular. In the near future, implanted medical devices, wearable electronics, and IoT will benefit most from STE-generated electricity. Powering personal electronics, implantable medical devices, and Internet of Things applications demands high thermoelectric efficiency, biocompatibility, and extremely varied thermal inputs.

These applications need lightweight operation and nonplanar substrate compatibility, not large-scale power generation. Power generation is less important than lightweight operation and nonplanar substrate compatibility for these applications. These applications need TEGs for devices of various sizes. Modern equipment and infrastructure make these procedures industrially viable.

Four major specifications for this application class are as follows:

1. Deliver enough electricity for the duty cycle,
2. Ensure power output without exhaustion,
3. Occupy an implantable form factor, and
4. Use only harmless materials.

Thermoelectric devices can run on body heat. TEG power is maintained by the temperature gradient, preventing the need for surgery. In unmanned space missions for decades, thermoelectric power generation has proven durable. Using carbon-based solar thermoelectric generator (STEG) devices makes flexible, biocompatible devices that can be implanted or worn safely and comfortably. STEGs need a 2°C temperature gradient to power life-saving medical implants. Thermal flux over the body's 1.5 m^2 surface is energy. At a 5°C thermal gradient, modern STE materials ($ZT \approx 1$) can generate around 20 μW/cm^2 from a human body. A pacemaker's power supply is around 10 μW, comparable to a STEG device with a surface area of 1 cm^2 for implanted devices.

STEGs with Seebeck coefficients of 100 μV/K and a 5°C temperature gradient require 20–250 p-/n-couples. This is challenging for a small implanted device with 500 pairs because traditional converters require a minimum potential of 500 mV. Under such constraints, material synthesis and device fabrication are essential. Solid-state processing and layer-by-layer deposition consume time and money. Additionally, intriguing is the use of solution-based printing's automation and scalability to produce multi-pair devices.

These initial findings show that solution-based printing can overcome new application device design issues. These are substantial advances, but the sector still struggles with device design and engineering. Besides implantable medical devices and biological sensors, tiny, low-power personal electronics have expanded rapidly. "Internet of Things" networks everyday objects for sensing, processing, communication, and energy management. This will spark the next semiconductor boom, say most market forecasts. Specifically, IoT is in the "early adoption" state. It must eliminate battery dependence in this paradigm due to the exponential number of sensor nodes. Wireless power transfer generates IoT electricity. The latest technique uses magnetic coils. Magnetic resonance coupling maximizes energy transfer and minimizes nonresonant object loss.

Wireless power transfer is inefficient and unsuitable for billions of IoT devices due to long-term electromagnetic field exposure and scalability issues. TEGs collect waste heat to power and recharge small gadgets seamlessly. To power IoT, thermoelectric technologies are being researched. Wearables use body heat for IoT power, while nonwearables use device waste. Wearables are popular in STE because body heat powers personal electronics. TEG wearables have only been sold as the Matrix Powerwatch. This smartwatch generates thermoelectricity from body heat. A STEG with a few cm^2 surface area can provide

20–30 μW of power for smartwatches. Besides smartwatches, wearable STEGs power portable music players, wireless headphones, and GPS devices. The STEG's surface area greatly lowers this value. Wearable thermoelectric technology cannot power smartphones because they use 1 W. This suggests a trend toward solving STEGs' biggest device and engineering problems. Wearable products can degrade thermoelectrically or fail structurally (delamination or cracking) with bending, folding, etc. Encapsulation prevents user exposure and performance loss.

The STEG module must diffuse heat and maintain a consistent temperature gradient for a stable power supply, indicating the need for soft heat exchangers, another neglected material. Developed heat exchanger materials are mostly hard inorganic. STEGs must efficiently dissipate resistive heat when connected to electrical devices like smartwatches. Contrary to TEGs, conformal thermoelectric devices can handle curves. IoT networks using everyday electrical goods require the form factor adaptability of STEGs. These applications struggle with heat unreliability. TEGs require consistent power and voltage.

TEG modules can continue to operate if one fails; hence, paralleling them may increase device life. These variables considerably enhance the form factor of such technologies, affecting materials development goals. Thermoelectric materials with low Seebeck coefficients (below 50 μV/K) are unsuitable for mixed TEG arrays, which require stable input voltages (20–500 mV) [25].

8.6 TEG MATERIALS AND DEVICES INFLUENCES ON CLIMATE CHANGE AND GLOBAL WARMING STRATEGIES

Understanding semiconducting polymer structure-property relations and potential composite components may optimize processing, which has various changing features. Customized organic semiconductors for thermoelectric purposes may lead to composite thermoelectrics using semiconducting polymers with specific CNT interactions.

Organic and hybrid organic–inorganic thermoelectrics are rare and efficient. Doping, thermal, and electrical transport mechanisms in complicated systems with several components and ordered and disordered phases make next-generation thermoelectric materials difficult to understand. The mechanical properties and power factor of novel thermoelectric materials and composites are prioritized. Innovative thermoelectric devices and applications should use mechanical flexibility to improve efficiency. Crystalline frameworks include gels, foams, conjugated microporous polymers, metal-organic, and covalent organic frameworks. Many intriguing material investigation paths make thermoelectric materials and devices promising [26].

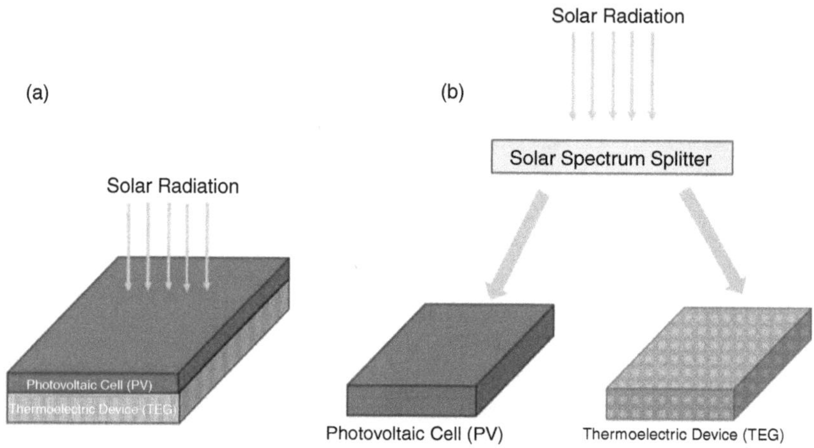

FIGURE 8.14 Schematic of the hybrid PV-TEG: (a) combined (PV-TEG) system and (b) spectrum splitting (PV-TEG) system [27].

Solar is the most common renewable energy source due to high radiation. Solar thermal modules and PV directly convert sunlight into heat and power. Solar PV's power density makes it overused in structures. Surface operating temperature cooling methods have increased PV system conversion efficiency. Photon waste heat below or above the solar cell band-gap is converted to power using thermoelectric modules to reduce PV surface temperatures. Recently, PV solar cell integration with thermoelectric modules has been intensively researched. PV and TE modules power the hybrid. Solar cell efficiency, longevity, and heating are improved by hybrid PV/TE systems. Spectrum splitting and combined PV–TE designs can optimize hybrid PV/TE systems with cooling systems (Figure 8.14).

In Figure 8.14a, parallel PV and TEG modules absorb UV and convert sunlight to energy. The TEG module heats the hot side by absorbing the remaining IR energy. A heat sink at the bottom of the integrated system can maintain a thermoelectric module temperature gradient. (b) The PV cell absorbs solar energy below the cut-off wavelength and reflects it to the thermoelectric device above it (Figure 8.14). Researchers have enhanced PV–TE systems in various ways. Hybrid PV–TEG systems outperform standalone PV systems under the same conditions. The efficiency increment shows that a design or situation can boost efficiency by over 30%. By adjusting PV cell and TEG module materials, operating conditions, and designs, a hybrid PV–TEG system can maximize conversion efficiency.

Biomass-burned chemical energy is a promising and reliable heat source for power, heating, agriculture, and more. Biofuel competes with coal, petroleum, and nuclear fuel as an environmentally friendly recycled energy source. Biomass energy is more weather-dependent than solar or wind. Biomass power plants face economic, capacity, and placement difficulties. Biomass energy system studies

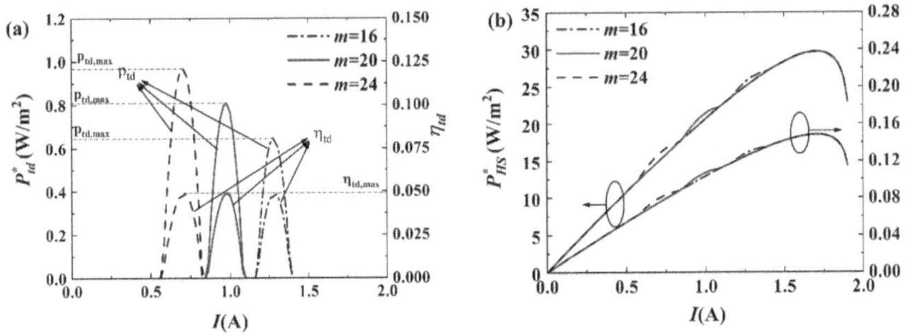

FIGURE 8.15 (a) TEG-TEC and (b) hybrid system performance at different m [28].

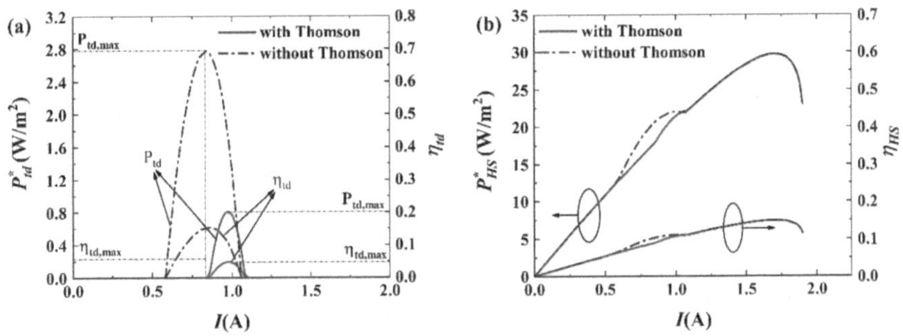

FIGURE 8.16 Effects of the Thomson effect on the performance of: (a) TEG-TEC and (b) hybrid system [28].

examine design, energy conversion system size, and hybridization with other renewable energy systems [27].

TEG–TEC uses m thermoelectric elements. Figure 8.15a shows that TEG–TEC's operating current range and maximum power density increase with m (the number of thermoelectric pairs in the TEG), while its efficiency remains approximately the same. The m influences hybrid system performance solely within TEG–TEC's working current range (Figure 8.15b), unlike other coupling system characteristics.

In Figure 8.16a, the Thomson effect decreases P* td and, reducing the TEG–TEC current range. Thomson effect lowers coupled system operating current range and performance (Figure 8.16b).

A linked system creates broad-spectrum solar electricity and cools. Evaluate the main irreversible losses to determine the relationship between PVM and TEG–TEC. Comparing the hybrid system to a single PVM module in its operational range, numerical calculations show a 3.97% increase in energy efficiency and a 3.98% increase in power density. Increased solar irradiation and diode ideality factor improve coupled system performance. High ambient temperatures enhance system performance [28].

Solar-thermoelectric-cooled ceilings (STCC) replace hydronic radiant panels. Solar thermoelectric cooled ceilings reduce cooling demand. Aluminum radiant panels and heat pipe sinks sandwich series-connected TE modules in the STCC. Fan-forced air convection helps TE modules dissipate heat. Controlling current direction cools and heats. A thermoelectric ventilation system dehumidifies fresh air in the combined system. A thermoelectric ceiling was tested for cooling and heating. Operating voltage increased heat flux. Reducing the ambient-to-indoor temperature differential increased heat flux and system COP in both cooling and heating modes. STCC's cooling mode heat flux exceeded 60 W/m^2, and COP achieved 0.9 at 5 V. In heating mode, the STCC system's COP hit 1.9, and its heat flux exceeded 110 W/m^2 at 4 V [29].

Global power usage reached 25,000 billion kWh in 2018. Hydroelectricity, nuclear, thermal, solar, and other sources supply the world's vast electricity needs. Industrial processes (30%–40%), internal combustion engines (25%), and solar systems (15%) generate electricity. This method uses 35% of the electricity, with the rest as waste heat. Since the population is rising daily, scientists are studying renewable, eco-friendly, and healthy sources. Researchers discovered thermoelectric energy while investigating the aforementioned statement, as discussed in the historical prospect.

Powering specific applications requires maximum thermoelectric energy utilization. This requires PCMs and efficient thermal-to-electricity generators, which increase power generation. Because of this, PCMs are becoming the backbone of many industries and growing exponentially. Major companies that research, apply, and commercialize daily PCM-based goods are shown in Figure 8.17a and b. Beginner PCM scholars may find this figure useful.

8.7 HIGH ZT IS NOT THE ONLY CONCERN

The entire TE community has focused on phonon and electron engineering to increase ZT, especially peak ZT. However, a high peak ZT within a limited temperature range won't increase efficiency for practical applications. In large heat sources, like solar heat or low-cost heat sources, power density is as important as efficiency (Figure 8.18).

However, decreasing leg length can have the following severe consequences: (a) an increase in large heat flux, which will raise heat management costs at the cold end; (b) an increase in the ratio Rc/R, which increases parasitic loss and decreases output power generation; and (c) an increase in thermal shear stress, which can cause device failure. Thus, increasing the TE material's power factor (PF) and optimizing its physical design parameters is optimal. As a pure material parameter, PF can be used to identify new TE materials with high output power.

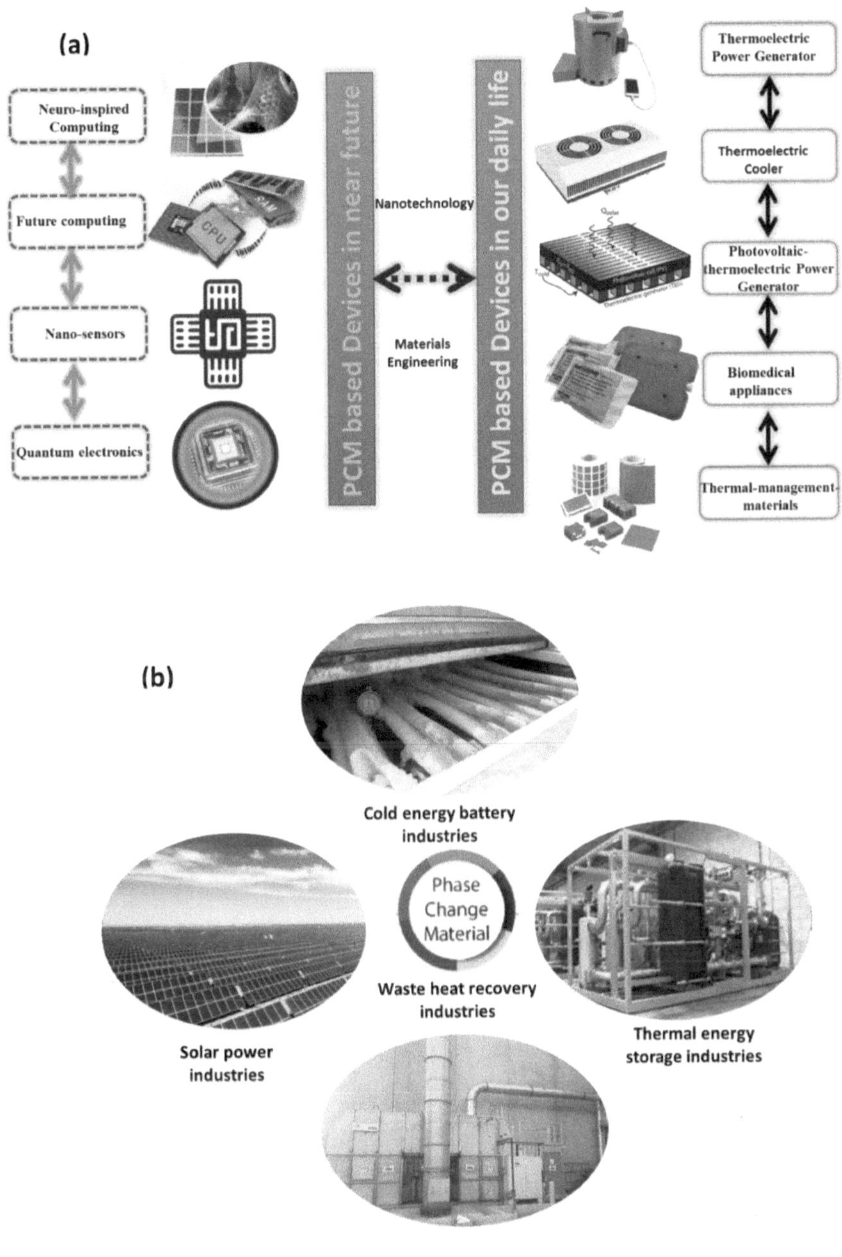

FIGURE 8.17 (a) Diagram of the most popular PCM-based gadgets and products for daily usage. Interdisciplinary PCM research worldwide will make advanced technology and devices available in 10 years. (b) Four important modern heavy industries who innovate and fabricate innovative PCMs [30].

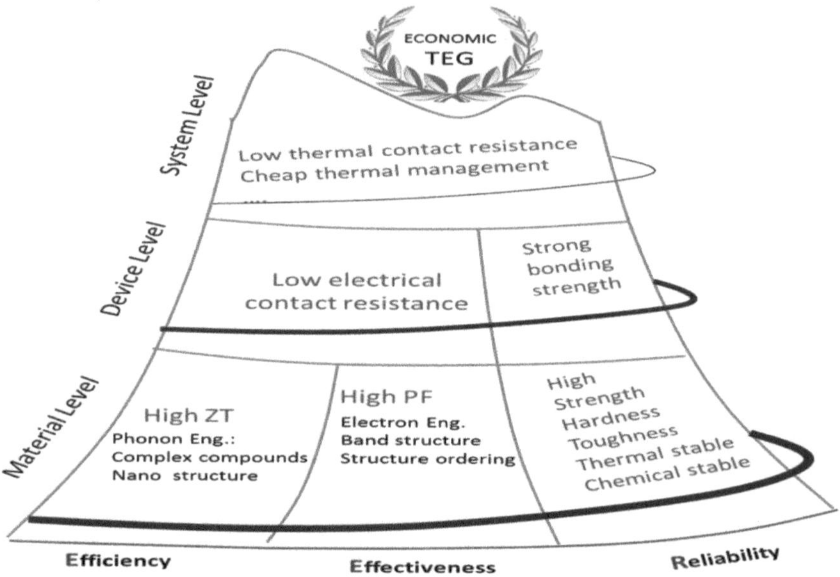

FIGURE 8.18 The efficiency–effectiveness–reliability mountain of TEG requirements [31].

REFERENCES

1. Rawat, P., Lee, M. H., Kumar, A., Thoravat, S., & Rhyee, J. S. (2024). Thermoelectric properties of extrinsic phase mixing in chalcogenide bulk nanocomposites. *Current Applied Physics, 60*, 15–31.
2. Zhang, D., Ramiah, J., Cagirici, M., Saglik, K., Solco, S. F. D., Cao, J., ... & Suwardi, A. (2024). Thermoelectric nanowires for dense 3D printed architectures. *Materials Horizons, 11*, 847–854.
3. d'Angelo, M., Galassi, C., & Lecis, N. (2023). Thermoelectric materials and applications: a review. *Energies, 16*(17), 6409.
4. Tohidi, F., Holagh, S. G., & Chitsaz, A. (2022). Thermoelectric generators: a comprehensive review of characteristics and applications. *Applied Thermal Engineering, 201*, 117793.
5. McGrail, B. T., Sehirlioglu, A., & Pentzer, E. (2015). Polymer composites for thermoelectric applications. *Angewandte Chemie International Edition, 54*(6), 1710–1723.
6. LeBlanc, S., Yee, S. K., Scullin, M. L., Dames, C., & Goodson, K. E. (2014). Material and manufacturing cost considerations for thermoelectrics. *Renewable and Sustainable Energy Reviews, 32*, 313–327.
7. Ouyang, Z., & Li, D. (2018). Design of segmented high-performance thermoelectric generators with cost in consideration. *Applied Energy, 221*, 112–121.
8. Patil, D. S., Arakerimath, R. R., & Walke, P. V. (2018). Thermoelectric materials and heat exchangers for power generation – a review. *Renewable and Sustainable Energy Reviews, 95*, 1–22.
9. Wang, W., Thesiya, D., & Mukhopadhyay, S. M. (2024). Hierarchical hybrid heat sink material for thermo-electric generators. *Applied Thermal Engineering, 236*, 121674.

10. Du, K. W., & Wu, C. I. (2024). An innovative tubular thermoelectric generator (TTEG) for enhanced waste heat recovery in industrial and automotive applications. *Applied Sciences*, *14*(2), 685.
11. Feng, T., Zhou, Z., Wang, P., Liao, Z., Wang, Y., Zhao, H., & Liu, W. (2024). Transverse thermoelectric materials: recent advances and challenges. *Next Energy*, *3*, 100105.
12. Ojha, A., Sabat, R. K., & Bathula, S. (2024). Advancement in half-Heusler thermoelectric materials and strategies to enhance the thermoelectric performance. *Materials Science in Semiconductor Processing*, *171*, 107996.
13. Kim, M., Park, D., Su, P. C., & Kim, J. (2024). Enhancing thermoelectric performance through acid-treated NbFeSb: a versatile approach for sustainable energy conversion. *Journal of Alloys and Compounds*, *976*, 173243.
14. Liu, Q., Huang, T., Chen, X., Gu, S., Duan, T., Lu, M., & Lin, J. (2024). Measurement of non-equilibrium characteristics of thermoelectric materials. *Journal of Materials Research and Technology*, *28*, 2740–2751.
15. Acharya, A., Chanda, B., Saminathan, M., Perumal, S., Jayanthi, K., Annapurna, K., ... & Mishra, S. K. (2024). Influence of metal organic framework glasses on thermoelectric properties of $AgSb_{0.96}Zn_{0.04}Te_2$ alloy. *Journal of Non-Crystalline Solids*, *627*, 122816.
16. Awasthi, C., Meena, R., Kandasami, A., & Islam, S. S. (2024). Structural and thermoelectric properties of $MoSe_2$/CNT nanocomposites. *Journal of Physics and Chemistry of Solids*, *184*, 111726.
17. Mostafa, S., Besisa, D. H., Zahran, F., Ahmed, A., Elwan, M., Khalifa, A., & Ewais, E. M. (2024). Processing and enhancement of thermoelectric performance of Nb doped ZnO (NZO) ceramics. *Ceramics International*, *50*(4), 6557–6566.
18. Hassan, M., Rouf, S. A., Alofi, A. S., Mahmood, Q., Ishfaq, A., mana AL-Anazy, M., ... & SayedYousef, E. (2024). Impact of terbium (Tb) on ferromagnetism and thermoelectric behaviour of spinels $MgTb_2$ $(S/Se)_4$ for spintronic. *Materials Science and Engineering: B*, *300*, 117110.
19. Twaha, S., Zhu, J., Yan, Y., & Li, B. (2016). A comprehensive review of thermoelectric technology: materials, applications, modelling and performance improvement. *Renewable and Sustainable Energy Reviews*, *65*, 698–726.
20. Chang, Y., Huang, Y. H., Lin, P. S., Hong, S. H., Tung, S. H., & Liu, C. L. (2024). Enhanced electrical conductivity and mechanical properties of stretchable thermoelectric generators formed by doped semiconducting polymer/elastomer blends. *ACS Applied Materials & Interfaces*, *16*, 3764–3777.
21. Liu, W., Hu, J., Zhang, S., Deng, M., Han, C. G., & Liu, Y. (2017). New trends, strategies and opportunities in thermoelectric materials: a perspective. *Materials Today Physics*, *1*, 50–60.
22. Jahn, W. (2021). Performance assessment of thermoelectric self-cooling systems for electronic devices. *Applied Thermal Engineering*, *193*, 117020.
23. Fleurial, J. P. (2009). Thermoelectric power generation materials: technology and application opportunities. *JoM*, *61*, 79–85.
24. Pourkiaei, S. M., Ahmadi, M. H., Sadeghzadeh, M., Moosavi, S., Pourfayaz, F., Chen, L., ... & Kumar, R. (2019). Thermoelectric cooler and thermoelectric generator devices: a review of present and potential applications, modeling and materials. *Energy*, *186*, 115849.
25. Zaia, E. W., Gordon, M. P., Yuan, P., & Urban, J. J. (2019). Progress and perspective: soft thermoelectric materials for wearable and Internet-of-Things applications. *Advanced Electronic Materials*, *5*(11), 1800823.

26. Finn, P. A., Asker, C., Wan, K., Bilotti, E., Fenwick, O., & Nielsen, C. B. (2021). Thermoelectric materials: current status and future challenges. *Frontiers in Electronic Materials*, *1*, 677845.
27. Wehbi, Z., Taher, R., Faraj, J., Castelain, C., & Khaled, M. (2022). Hybrid thermoelectric generators-renewable energy systems: a short review on recent developments. *Energy Reports*, *8*, 1361–1370.
28. Zhang, X., Huang, Y., & Chen, Z. (2022). A hybrid system integrating photovoltaic module and thermoelectric devices for power and cooling cogeneration. *Solar Energy*, *239*, 350–358.
29. Contento, G., Lorenzi, B., Rizzo, A., & Narducci, D. (2017). Efficiency enhancement of a-Si and CZTS solar cells using different thermoelectric hybridization strategies. *Energy*, *131*, 230–238.
30. Pandey, R., Thapa, P., Kumar, V., Zhu, Y., Wang, N., Bystrzejewski, M., & Tiwari, S. K. (2022). Updates in phase change materials for thermoelectric devices: status and challenges. *Materialia*, *21*, 101357.
31. Liu, W., Jie, Q., Kim, H. S., & Ren, Z. (2015). Current progress and future challenges in thermoelectric power generation: from materials to devices. *Acta Materialia*, *87*, 357–376.

Index

FSC
www.fsc.org
MIX
Papier | Fördert
gute Waldnutzung
FSC® C083411

Kolibri 360 GmbH
Ferdinand-Jühlke-Straße 7
99095 Erfurt, Deutschland
produktsicherheit@kolibri360.de